Post-Pandemic Sustainable Tourism Management

Tourism, as with many parts of the economy, is at a pause-reflect-rest stage in the post-pandemic world. This book puts forward some positive and practical concepts for the reset stage in terms of pushing towards wholly sustainable tourism.

The COVID-19 pandemic has been disastrous in terms of the loss of human life, the physical and mental strains placed on large numbers of populations across the globe who have been quarantined in their homes and in terms of the costs of dealing with the pandemic and supporting business and citizens through the period. Tourism has been comprehensively damaged, not only in advanced economies, but also in poorer developing economies where tourism provides a vital source of income and employment. The problem has been complicated by the shattering effect on mass tourism, which has been far more sensitive to the shutdown of travel and accommodation than ethical and responsible tourism activities focused at a local sustainable level. Therefore this book evaluates how the pandemic and economic decline affects ethical and responsible tourism – the type of tourism which sustains and develops local communities in a balanced way for the benefit of future generations. It reflects on the position the authors established in *Ethical and Responsible Tourism: Managing Sustainability in Local Tourism Destinations* and then determines how ethically and responsibly focused tourism may adapt, develop and maintain safety for consumers in the post-virus world.

This book will be essential reading for students, researchers and practitioners of tourism, environmental and sustainability studies.

Marko Koščak is currently an Associate Professor at the University of Maribor, Faculty of Tourism Brežice, Slovenia. In the past 30 years, he has worked as an Advisor to UNDP LoSD on sustainable tourism initiatives in the South East Balkan countries of Croatia, Serbia and Montenegro, Bosnia, Kosovo and Macedonia. He was also involved in a number of sustainable tourism projects and development initiatives across Europe and Asia. He is the Founding Member of the Slovenian Rural Development Network, which is part of the European Network "PREPARE".

Tony O'Rourke works with Green Lines Institute for Sustainable Development in Portugal. Since retiring in 2011, he has taught part-time at MSc and MBA level as a visiting/part-time Professor and also carried out advisory work for Co-operatives UK. His current academic interests are in the fields of Sustainable and Responsible Local Tourism, Financing of Local Tourism/Destination Management organisations and the creation of tourism credit co-operatives.

Routledge Focus on Environment and Sustainability

The Emerging Global Consensus on Climate Change and Human Mobility
Mostafa M Naser

Traditional Ecological Knowledge in Georgia
A Short History of the Caucasus
Zaal Kikvidze

Traditional Ecological Knowledge and Global Pandemics
Biodiversity and Planetary Health Beyond Covid-19
Ngozi Finette Unuigbe

Climate Diplomacy and Emerging Economies
India as a Case Study
Dhanasree Jayaram

Linking the European Union Emissions Trading System
Political Drivers and Barriers
Charlotte Unger

Post-Pandemic Sustainable Tourism Management
The New Reality of Managing Ethical and Responsible Tourism
Tony O'Rourke and Marko Koščak

Consumption Corridors
Living a Good Life within Sustainable Limits
Doris Fuchs, Marlyne Sahakian, Tobias Gumbert, Antonietta Di Giulio, Michael Maniates, Sylvia Lorek and Antonia Graf

For more information about this series, please visit: www.routledge.com/
Routledge-Focus-on-Environment-and-Sustainability/book-series/RFES

Post-Pandemic Sustainable Tourism Management

The New Reality of Managing Ethical and Responsible Tourism

Marko Koščak and Tony O'Rourke

Routledge
Taylor & Francis Group

LONDON AND NEW YORK

First published 2021
by Routledge
2 Park Square, Milton Park, Abingdon, Oxon OX14 4RN

and by Routledge
605 Third Avenue, New York, NY 10158

Routledge is an imprint of the Taylor & Francis Group, an informa business

British Library Cataloguing-in-Publication Data
A catalogue record for this book is available from the British Library

Library of Congress Cataloging-in-Publication Data
A catalog record has been requested for this book

ISBN: 978-0-367-71626-4 (hbk)
ISBN: 978-0-367-71631-8 (pbk)
ISBN: 978-1-003-15310-8 (ebk)

Typeset in Times New Roman
by Taylor & Francis Books

Contents

List of tables	viii
Foreword	x
List of contributors	xiii
Acknowledgements	xv
Abbreviations	xvi
Introduction: The new reality	1
1 The pre-pandemic situation	3
2 A new tourism world	13
3 The economic and financial consequences for tourism	27
4 Some Reflections	48
5 New dimensions in a post-COVID world	55
6 The local tourism perspective	68
7 The Case Studies	80
8 Analysis and conclusions	133
Index	145

Tables

2.1	COVID-19 cumulative data at 31 December 2020 – top five EU/EEA countries	14
2.2	COVID-19 effects on five high value EU tourism economies	17
2.3	UNWTO forecasts – international tourist arrivals	17
3.1	Global, EuroZone and G7 GDP (% change)	28
3.2	Composite Leading Indicators* (main global economies). (Month-on-month % change (m-o-m) for Apr/Jun/Aug 2020 and year-on-year change (y-o-y) for Aug 2020/Aug 2019)	29
3.3	Tourism as an economic feature	32
3.4	Tourism and the labour market	33
3.5	GDP changes in EU's largest tourist economies	33
3.6	Unemployment increases in EU's largest affected labour markets (based on seasonally adjusted data)	34
3.7	Domestic tourism – OECD's five largest domestic markets	40
3.8	Domestic tourism – OECD's five smallest domestic markets	41
CS1.1	Index of tourist visitor growth 2015–2016 by country	88
CS2	Vineyard tourism statistics (2016–2019)	94
CS3.1	Tourism expenditure in Italy (2018)	99
CS3.2	Tourism expenditure in EU27 (2018)	100
CS3.3	Italy – share of EU27 tourism expenditure (2018)	100
CS3.4	The impact of COVID-19 on tourism in Italy	101
CS3.5	GDP quarterly percentage change Eurozone/Italy	102
CS3.6	Current Account Balance as percentage of GDP	102
CS3.7	Density of day tourism	103
CS3.8	Progress of COVID-19 in Provincia di Venezia (total infections on date)	107

CS4.1   Inbound tourists 2017–2019   110
CS4.2   Tourist expenditure 2017–2019   110
CS4.3   Tourist expenditure 2017–2019 per capita   111
CS4.4   Origin of tourists 2019 – top 5 originating countries   112
CS4.5   Tourism expenditure per capita – top spending
        countries (Jan. to June 2020)   114
CS4.6   Tourism expenditure per capita – top countries by
        number of tourists (Jan. to June 2020)   114
CS4.7   Malta's current account balance as percentage of GDP
        (with comparison to the EuroZone)   115
CS4.8   Malta's Exports of Goods & Services Q1 2018–2020   115
CS4.9   Malta's real GDP growth 2019–2021 (with comparison
        to the EuroZone)   116
CS4.10 Malta's real GDP growth Q1 2018–2020 year-on-year
        (with comparison to the EuroZone)   116
CS5.1   Number of visitors by transport mode in Logarska
        Valley, 2017–2020   123
CS5.2   Analysis of visitors by transport mode in Logarska
        Valley (2019)   124
CS5.3   Updated analysis for Logarska Valley (May–September
        2020)   124

Foreword

This book follows on from research conducted by the authors since 2014 into many elements of responsible and ethical tourism manifested at the local level. This resulted in *Ethical and Responsible Tourism: Managing Sustainability in Local Tourism Destinations*, published by Routledge in November 2019 (ISBN: 978-0-367-19144-3 (hardback); ISBN: 978-0-367-19146-7 (paperback) and ISBN: 978-0-429-20069-4 (e-book). Information from: https://www.routledge.com/Ethical-and-Responsible-Tourism-Managing-Sustainability-in-Local-Tourism/ORourke-Koscak/p/book/9780367191467).

In this co-edited text-book, we along with 38 other contributors, have outlined the position in regard to the management of responsible tourism at a local level, and reflected on the situation as it existed in the latter part of 2019. This included the important value of tourism which resisted the destruction of fragile local environments (urban or rural); the ability of local stakeholders to develop a form of tourism which met capacity requirements and the concept of protecting the physical and cultural heritage whilst also ensuring a degree of socio-economic benefit to local communities.

However, none of us engaged with that publication had at that time a concept of the catastrophic COVID-19 pandemic which emerged in the early months of 2020. By March 2020, global travel and accommodation systems had been effectively closed and the tourism industry was in suspension. Mass tourism destinations had become silent and empty; millions of employees whose livelihood depended on travel and tourism had their employment suspended or were only receiving a portion of their previous salaries. Large travel companies and airlines have become dependent on state loans or bailouts to remain in business, whilst many smaller travel and tourism companies have collapsed into insolvency. The COVID-19 pandemic has been disastrous – loss of life as well as physical and mental strains on populations. In the autumn of 2020, despite the widespread lockdowns earlier in the year, we have seen the occurrence of second spikes requiring further restrictive measures.

At the same time a global economic depression is developing which may be far greater in reach than the global financial crash and economic depression of 1929–1935 and consequential on this have been increasingly higher levels of debt in advanced economies. Societal impacts are also to be taken account of, as Antonio Alberto Clemente (Antonio Clemente (2020) "Le Covidor" in Bilò, F., Palma, P. (ed.) *Il cielo in trentatré stanze - cronache di architetti #restati a casa*, pp.38–39. Lettera Ventidue: Siracusa) states:

> The virus created an altogether peculiar condition: man has become an island and society, an archipelago in which the only relationship of reciprocity has been the visual-sound one, always provided that the interpersonal safety distance was respected. Born as a temporary solution, social distancing will continue to be practiced voluntarily in every city, even after the end of the emergency phase.
>
> Concerning the anguish generated by the intrinsic characteristics of the virus, it cannot be found due to material nonexistence; but it is there. COVID-19 is not a determined object which one can defend against: it is invisible but present. This means that it can be anywhere. A lethal atopy that can lurk everywhere and potentially in anyone who crosses the city. Hence the consequence that the best solution could be introversion.
>
> The imprisonment in oneself to avoid any physical contact; distancing no longer as an obligation but as a choice. If things were to go in this direction then COVID-19 could even undermine the Aristotelian assumption of man as a social animal.

Therefore we are now considering as to how both the pandemic and the resultant economic issues have and will affect ethical and responsible tourism – that type of tourism which sustains and develops local communities in a balanced way for the benefit of future generations. It was developed as a tourism model which we saw as being the way forward for developing ethical and responsible tourism at a local level. The concept of this new publication is therefore to focus on the position we established in *Ethical and Responsible Tourism: Managing Sustainability in Local Tourism Destinations* and then to determine how ethically and responsibly focused tourism is likely to adapt, develop and hopefully to succeed in this new reality. We will consider how we are required to pause, reflect and then to reset our under-standing and objectives.

We believe that it is important to make a clear connection to the previous text-book; accordingly we are including identifiable links into *Ethical and Responsible Tourism: Managing Sustainability in Local Tourism Destinations*. The links will cite chapter/page number for elements which are common and relevant to both publications; in addition we are continuing the concept of including further reading and questions at the end of each chapter. This will ensure that this book can be useful for both teaching and training purposes, as well as illuminating and enlarging this highly topical subject area as we move into the new reality for sustainable tourism.

We remain committed to the concept that local problems require local solutions, nonetheless we also understand that local destination issues require an interdisciplinary approach and the contribution of different disciplines and expertise in resolving development challenges. This has become a pressing imperative as mass tourism continues to stagger from the loss of markets and its inability to cope with the protective and safety structures that living with COVID-19 have now imposed on tourism activity. Locally based sustainable tourism, as a flexible and sustainable product may appear to have a much brighter prospect over the medium to longer term as well as converging with the reality of a more sustainable planet and the reduction of excess consumption.

Contributors

Marko Koščak held the position of Assistant Professor from 2014 to 2019 and is currently an Associate Professor at the University of Maribor, Faculty of Tourism Brežice, Slovenia. He studied in Ljubljana (Slovenia), Birmingham (UK), Vienna (Austria) and in 1999 completed his PhD on the thesis "Transformation of Rural Areas along the Slovene – Croatian Border" at the Faculty of Arts, Department of Geography, University of Ljubljana. His academic interests are in the field of Sustainable and Community Tourism, Geography of Tourism and Destination Management. These topics are his research work interest fields in which he also lectures to students. His professional work career started on different activities in the field of Rural Development in Slovenia and abroad since 1986, when he commenced implementation of Integrated Rural Development Projects on local-community level. He was a Project Manager of the Dolenjska and Bela krajina regional sustainable tourism initiative Heritage Trails in South East Slovenia from 1996 to 2009, under the umbrella of the Chamber of Commerce Novo mesto. Since 1986 he was also a Regular Consultant with the Ministry of Agriculture, and employed there from 1999 to 2001 as an Advisor to the Slovenian government in the Sector for Structural Policy and Rural Development. In the past 30 years, he has worked as an Advisor to UNDP LoSD and sustainable tourism initiative in South East Balkan countries of Croatia, Serbia and Montenegro, Bosnia, Kosovo and Macedonia. He was also involved in a number of sustainable tourism projects and development initiatives and worked on these in Europe and Asia. He is the Founding Member of the Slovenian Rural Development Network, which is part of the European Network "PREPARE". His professional expertise and experiences are primarily in the following fields: Sustainable and Responsible Community Development, Rural and Eco Tourism, Economic

Diversification on farms, Product Development, Heritage Tourism, Regional and Rural Development and Cross-border Cooperation.

Tony O'Rourke has been actively collaborating with Marko Koščak for a number of years in the area of sustainable and local tourism development. He studied at UK universities in Warwick, Edinburgh and Stirling. His current academic interests are in the fields of Sustainable and Responsible Local Tourism, Financing of Local Tourism/Destination Management organisations and the creation of tourism credit co-operatives. His professional work career began in Higher Education Management connected to training and resources application. From 1990 to 2001 he was working with Scottish Financial Enterprise, the University of Stirling, Heriot-Watt University and the European Commission DG23 on small business financing at a micro level as well as the financing of tourism enterprises in Central and Eastern European transition economies. This included monitoring EU projects in Czech Republic, Hungary, Kazakhstan, Poland and Slovenia. From 1996 to 2001, he was Secretary-General of the Association of European Regional Financial Centres, and from 1992 to 2004 he was Professor in small business financing/small financial market strategy at universities in Ireland, Montenegro and Serbia, and at the same time an Expert Advisor to Dun & Bradstreet Country Risk Services on Bosnia & Herzegovina and Serbia. In 2004, he returned to the University of Stirling as the Director of the programme for continuing professional development in the Scottish finance and investment sector, as well as teaching on the MSc Banking and Finance Programme. Since retiring in 2011, he has taught part-time at MSc and MBA level as a visiting/part-time Professor and also carried out advisory work for Co-operatives UK. He is currently working with Green Lines Institute for Sustainable Development in Portugal.

Acknowledgements

The authors would wish to take this opportunity to thank first of all our families, who have had to endure the effects of the work in which we have both been involved in the process of creating this book. It has taken much effort and time, and we appreciate their willingness to give us the space to complete this task.

So therefore very grateful thanks to Branka and children as well as to Elizabeth for their continuing support, patience and resilience. COVID-19 has had a huge effect on our family life and the way in which we live.

However, to some extent it has provided both of us with the space and time to consider its impact on sustainable tourism. Hopefully the new reality of the tourist world will enable us to focus on longer-term objectives and concepts which will support the post-COVID world in which we and future generations will live and hopefully flourish.

We also wish to thank all the editorial, publishing and marketing staff at Routledge, especially Fran and Rosie, who have assisted us in this latest project, particularly through the difficult times during the initial and subsequent COVID-19 lockdowns.

Marko Koščak and Tony O'Rourke

Abbreviations

Currencies quoted

EUR	Euro
USD	US dollar
GBP	UK pound
AUD	Australian dollar

Institutions

EU	The 27 member states of the European Union
EEA	European Economic Area (EU+Norway, Iceland & Liechtenstein)
IMF	International Monetary Fund
OECD	Organisation for Economic Co-operation & Development
UNWTO	United Nations World Tourism Organization

Economic indicators

GDP	Gross Domestic Product
BoP	Balance of Payments
C/A	Current Account balance

Currency values

Due to the current variations and levels of currency instability, we have given currency values in the appropriate national/international denominator (e.g. EUR/USD/GBP/AUD), without providing exchange rates.

However to give a general guide to currency values please find below the average inter-bank rates for the relevant currencies for the period from 01.03.20 to 16.11.20.

	EUR	*USD*	*GBP*	*AUD*
EUR	*1.0000*	1.1873	0.9019	1.6577
USD	0.8419	*1.0000*	0.7738	1.4089
GBP	1.1070	1.3034	*1.0000*	1.8342
AUD	0.6112	0.7260	0.5514	*1.0000*

As indicated on page, we are providing links between this publication and our previous publication. This relates to the significant changes in sustainable tourism over the period from January 2020; the tourism world changed substantially, but we believe it is important to connect the ideas and concepts which we were proposing at the end of 2019 to the situation which existed by the end of 2020. The links are therefore indicated as LINK, followed by **Chapter** and **page number** in the following publication:

Koščak, M. and O'Rourke, T. (2020) Ethical and Responsible Tourism: Managing Sustainability in Local Tourism Destinations. Routledge, Abingdon, UK & New York, USA.

Introduction
The new reality

To achieve sustainable levels of tourism requires management, marketing, promotional development, product and capacity development to be wholly integrated with an ethical and responsible approach. Whilst traditionally, tourism has sought to maximise people flows, ethical and responsible tourism seeks to regulate flows to ensure that footfall does not damage unique heritage and cultures whilst preserving historical and physical environments. Thus responsibility in tourism is concerned with creating and agreeing a balance between socio-economic objectives which benefit communities and the preservation of the human, cultural and physical environment.

The COVID-19 pandemic has been disastrous in itself in terms of the loss of human life as well as the physical and mental strains placed on large populations across the globe who continue to be quarantined in their homes. The pandemic first appeared in late December 2019 in China and then moved into neighbouring countries in SE Asia through January/February 2020, before reaching Europe in February/March 2020. Indications are that it may well continue into 2021 in the more advanced economies and potentially into 2022 in less under-developed economies. Added to this, governments in the advanced economies are having to meet the costs of dealing with the pandemic and supporting business and citizens at significant costs to their economies. Both virus and consequent costs summon the prospect of a global economic depression far greater in reach than that resulting from the 2007–2009 global financial crash and the economic depression of 1929–1935.

Therefore we consider how both the pandemic and resultant economic issues have and will affect tourism, and specifically ethical and responsible tourism – that type of tourism which sustains and develops local communities in a balanced way for the benefit of future generations. It was developed as a model which by the end of 2019, we saw as being the way forward for sustainability at a local level.

The concept of this new book is in part to reflect on our previous views – how we successfully manage sustainability in local tourism destinations – and then to determine how ethically and responsibly focused tourism is likely to adapt, develop and hopefully to succeed in the post-virus world. Whilst we optimistically describe the post-COVID world as the "new reality", at this point in time we have no substantive concept of how that reality may or may not evolve. Perhaps the following pages will pose questions that may assist all of us in answering that fundamental issue.

1   The pre-pandemic situation

Our previous publication in its conclusions reflected on issues affecting ethical and responsible tourism, and on its future development within the framework of the local tourism environment. The reflections focused on three important conceptual and practical areas as a guide:

- Three key themes of ethical and responsible tourism development – destination management aspects; environmental and social aspects and the business aspects
- Critical issues in the development of ethical and sustainable tourism in local destinations
- Understanding the longer-term view

These are described below.

The key themes

Destination management aspects

Destination management [*LINK: Chapter 13/pp. 188–189*] – successful performance requires a new and practical tourism paradigm combining excellence, co-creation and co-operation, and high quality services. Innovative product development concerns increasing competitiveness, facilitating sustainable tourism growth and consequently increasing tourism turnover. A systematic approach is important to achieve organisational synergy, creating new jobs, developing new skills and ecological innovations. This takes into account the competitive adjustments required for sustainable development.

Responsible tourism and sustainability – responsible tourism concerns the use of tourism to create improved places for people to live or to visit; sustainable tourism is focused on key players maximising positive socio-economic and environmental impacts of tourism at a local level. In effect responsible tourism drives progress towards sustainability (ICRT, 2007)

Participatory planning – effective collaborative planning depends on adequate representation of interests, a shared vision, goal accomplishment, good working relationships and open communication between stakeholders, requiring strong leaders and administrative support. In the first instance, participatory planning forms an integral part of destination management processes by capturing the importance of the public perspective in the procedure; secondly, by stressing the importance of the creative participatory methods to attract stakeholders and enhance their willingness to partake in the process; thirdly, by identifying creative participatory planning tools that may be applied to enhance participatory planning within the given procedures.

Carrying capacity – the carrying capacity study is necessary in order to identify environmentally and culturally sensitive areas and ensure that the tourism destination is sustainable. The purpose is to ensure that all tourists attracted to the particular destination will not have a deleterious impact on the cultural or natural sites, that overcrowding will not result in visitor dissatisfaction, and that local people will not feel antagonistic towards their "guests". This is essential if tourism is to contribute to the conservation of cultural and natural heritage though the realisation of economic value and raising awareness of, and commitment to, the local patrimony. Local people must be consulted in the assessment of landscapes and the heritage assets. It is essential to ensure that the local impact of increased heritage tourism is within tourism development and marketing processes.

Community-led development – this approach turns traditional "top-down" development policy on its head. Under community-led development, local people form and manage a local partnership to design and implement an integrated development strategy – the "bottom-up" approach. The strategy builds on the community's social, environmental and economic strengths or "assets" rather than simply compensating for its problems. In the phase of implementation of strategy, the partners receive long-term funding, and decide how it is spent.

Experiential and multi-sensorial – experiential travel is a form of tourism in which people focus on experiencing a country, city or particular place by connecting to its history, people and culture. Travellers can experience intimate encounters with real people and places without distraction. The focus is on travelling to less-travelled destinations while still providing a high-standard customer service. Respecting and focusing on the traveller is key – where they travel to and the inhabitants. In addition, it is shared opinion that sight, sound, smell, taste and touch are in a direct connection to the customers' emotions; therefore, it is crucial to understand and react to the emotional needs of customers (joy, awe, excitement, delight). Tourism experiences are increasingly determined by experience of co-creation and technology use.

Environmental and social aspects

Environment [*LINK: Chapter 13/pp. 189–190*] – tourism may be a lucrative source of revenue for a destination, but may also have major negative impacts on that destination. Whenever the negative impacts on the natural environment are dealt with, it should be considered that these rarely affect only one entity but ecosystems as a whole. The impacts on the natural environment not only affect pristine nature areas but also cultivated land – an important part of the natural and cultural heritage of a region, that is ecologically valuable because it is the habitat of many species. Environmental impacts at local and regional level may also affect the global environment in the long run (UNESCO, 2009).

Social responsibility – tourism aids change and development and thus has major effects on the cultural development of a society. The reactions of societies towards tourism are diverse: some reject changes; others inculcate them into their traditions; and some will abandon their cultural roots altogether. Whilst cultural change is an unavoidable, natural part of human culture, the sudden and forced changes that tourism often brings may cause the complete breakdown of a society and the loss of an entire cultural tradition. Socio-cultural impacts of tourism are often hard to identify or to measure; generally, tourism brings about changes in the value systems and behaviour of the people and causes changes in the communal structures, family relationships, collective traditional life styles, ceremonies and morality. The ambiguity of socio-cultural impacts is due to the fact that tourism may have effects that are beneficial for one group of a society but negative for others (UNESCO, 2009).

Local community – one motive for responsible travelling is the desire to interact with people in local destinations and understand their cultures. This exchange may support understanding, leading to the reduction of prejudices and contributing to decreasing inter-societal tensions. Tourists' demand for the original and authentic elements of a destination's culture may cause a revaluation of that local heritage and tradition, leading to a renaissance of indigenous cultures, arts and crafts and the rejuvenation of events that are becoming forgotten due to modern developments and adaptation to developed economy lifestyles (UNESCO, 2009)

Networking – to achieve ecological, socio-economically sustainable tourism, adequate management and monitoring requires to be established. Providing an optimal solution to such aims needs good networking between participant institutions and individuals. Importantly the different stakeholders involved in the tourism business are responsible for the implementation of multiple parts of the guiding principles. Governments, tourism businesses, local communities, NGOs, and tourists should all contribute in ensuring that tourism becomes more sustainable. In order to achieve sustainability the various actors should co-operate and stimulate each other to place the principles into practice and achieve this on all levels – international, national, regional and particularly local (UNESCO, 2009).

Climate change – many fragile and bio-diverse tourism destinations are particularly sensitive to the effects of climate change. This is not only apparent in regions where large-scale tourism appears to be on the increase but is also obvious in more temperate regions where volatile changes in rainfall as well as flooding and more dynamic storm patterns have become evident. Whilst tourism may have an impact on such fragile environments, climate change may also lead to a rapid withdrawal of tourism. We should therefore begin to view climate change as an issue on which tourism may have an effect or be affected by.

Renewable energy – it is clear that mass tourism has tended to be focused in the past on the use of non-renewable energy sources. Whilst this is changing, local tourism has the greatest ability to benefit from localised renewable power schemes. This includes the use of wind, solar and hydro power at a local community level. At the same time, we should be aware of the use of off-grid energy, water and waste consumption. Holiday homes in rural communities have the capacity to use alternative power sources for heating, grey water from rainfall for showers and bio-degradable technologies for the management of waste.

The business impacts

Business adaptation [*LINK: Chapter 13/p. 190*] – the material we have collected indicates a wide potential for assessing how businesses may embrace ethical and responsible development at a local level whilst maintaining financial stability. This should of course require looking at new models of business activity that have a strong base in local communities and to an extent focus on self-generation of community funding within an overall regenerative community concept.

Financing ethical and responsible tourism at a local level – local tourism activities will always be constrained by the availability of capital through traditional sources. There are however useful models for developing community, co-operative and mutual systems to finance local actions. Blending differing methodologies and looking at new ways to create financing structures is becoming an important feature, particularly if is it necessary to escape from total reliance on grant-aided models.

Investment through SMART technology – powerful growth in technology-enabled systems of financing has suggested the potential for accessing wider pools of investment opportunity for ethical and responsible tourism enterprises in peripheral regions. Traditionally peripheral and remote regions have suffered from a lack of direct access to financial markets – new technology, such as crowdfunding, enables a more direct and potentially sustainable approach.

Assessing market changes – tourism markets are in a continual process of change, and over the last three decades we have seen a number of shifts; whilst this has included a greater interest in responsible tourism and the reduction of the size of tourism groups to match footfall, at the same time, we have seen a dynamic growth in cruise tourism as the size of cruise vessels expands to meet market expectations.

Community-driven management – stakeholder engagement tends to bring in a mixture of public, private and social sector actors. Building on community engagement links the management and operation of local tourism destination organisations with the actual community. This then requires a review of ownership models and a potential for greater interest in mutual and co-operative structures which localise ownership but do provide access to improved management capacities and solutions.

Partnership – we should seek to distance ourselves from a simple dichotomy between tourism providers (individuals and organisations) and tourism recipients (consumers). We should also look at other agents in the community process; decision-making is often in the hands of the older members of society, whilst the future of ethical and responsible development will without doubt be in the hands of today's younger generation.

Critical issues in the development of ethical and sustainable tourism in local destinations

In *Ethical and Sustainable Tourism – Managing Sustainability in Local Tourism Destinations* [**LINK**: *Chapter 31/pp. 431–432*] we were able to determine a number of critical issues in the development of ethical and sustainable tourism in local destinations. These drew not only from the editorial material but also from the wide range of inputs and observations made by contributing authors to the book. We saw these critical factors as being in the form of "success" and alternatively "failure".

Critical success factors

1   The building of local coalitions of actors and agents is a primary success factor; however, it is also clear that there is a need to bring into the coalition elements of regional, national and international organisations to achieve the maximum effect.

2   Ethical and responsible event management, as a part of local tourism activity, appears to be successful not only in attracting tourism input, but also in engaging the support of local residents. Minimising waste from tourism is a potent example.

3   Promoting tourism all year round, particularly in peripheral regions which are not weather dependent, has a major economic benefit and, by spreading the tourism season over a wider base, has a less detrimental environmental effect.

4   Flexible and imaginative financial solutions, blending different funding methodologies (credit, equity and crowd funding finance), appear to be effective in financing local, ethical and responsible tourism development.

5   Tourism projects which have an environmentally friendly approach have the potential to push wider society into infrastructure developments which themselves articulate environmental responsibility and also will bring longer term socio-economic benefits.

6   Regeneration of local environments with appropriate educational developments and community engagement, potentially helps to secure and enhance the local culture and heritage. Local inhabitants who are aware and understanding of their own culture and heritage (and its future) will be more willing and receptive to the idea of sharing this unique inheritance with tourists.

7   Major urban heritage sites which are able to protect their unique physical and heritage environments, whilst at the same time offering inclusive experiences for all types of tourists (e.g. the mobile as well as the less mobile) may have a positive economic impact on local communities.

Critical failure factors

1   We would suggest that there are a number of capacity challenges at a local level which minimise the opportunity to develop sustainable activity. These include poor financial capacity, poor marketing and

promotional capacity and poor management capacity. These capacity problems are often due to a lack of co-operation between actors and agents as well as an unwillingness to create financial/management/marketing coalitions.

2 A serious failure is where local tourism actors and agents fail to create destination management structures; this failure then means that the opportunity to co-operatively and collectively promote and market a destination in a meaningful and targeted way is totally lost.

3 Overtourism remains a massive critical failure; it not only degrades the environment and escalates the living costs of local inhabitants, but it also creates antagonism between tourists and local residents. Furthermore the effects of overtourism do apply not just to major urban tourism destinations, but more increasingly to rural tourism centres as well.

4 Failing to understand the necessary balance between people and the environment – at whatever level – has the potential not only to diminish local culture and heritage but also to undermine economic stability and development. This is frequent and the result of failing to engage all sectors of society in the planning of tourism activity.

5 Tourism cannot be seen as a means of injecting a new format for economic growth into economies which have significant structural imbalances. In the short term, tourism may well provide economic growth and foreign exchange inputs; but tourism trends are tenuous and extremely affected by global economic and security trends. Today's favoured tourism destination may swiftly become abandoned and under-utilised tomorrow.

Understanding the longer-term view

We tend towards a view that the development of ethical and responsible tourism at the local level has been blighted and constrained by short-term thinking. Often this is driven by the blind acceptance of previously fashionable neo-liberal economic principles which predicated success based on contribution to GDP or significant gross income inflows. The evidence of the past two decades, particularly through the global economic crisis of 2007–2009 and the subsequent economic depression, indicates how short-term thinking by major financial and investment institutions brought about the near collapse of the advanced economies [*LINK: Chapter 31/pp. 432–434*].

Equally mass tourism – whether large-scale hotel developments or cruise ships of an increasingly massive size – are developments which are ethically challenging and potentially non-sustainable. Such developments, by the very capital required, may only be possible through the engagement of large-scale corporate investment which is dependent on hedge funds and private venture capital investors who will clearly not necessarily have an ethical and responsible outlook on their investment scenarios. At the same time, mass tourism is hugely sensitive to events and trends; the unwillingness of US tourists to travel after 9/11 was responsible for the collapse of a number of major airlines; we also saw a major shift of tourism flows from North Africa and Turkey after terrorist attacks in those regions. But it is clear from the contributions in this book that short-term economic inflows tend to benefit organisations, corporations and individuals outside local areas. The economic benefits of cruise tourism, for example, are primarily directed to international cruise companies and port facilities; they fail to provide immediate benefit to communities where tourism footfall damages and destroys both the tangible and intangible cultural and heritage environment. Clearly seeking to find a balance between the importance of economic growth (particularly in peripheral coastal or rural regions) and protection of fragile environments and cultures is a complex and frequently dangerous balancing act.

So how do we seek to determine the potential for meaningful longer-term planning? Perhaps this does involve looking at alternative planning perspectives which encourage ethical and responsible tourism at a local destination level. Involvement of stakeholders in strategic management of a destination is important in each stage of development. We would suggest that this emphasises the importance of a "bottom-up" approach in local tourism management. This involves growing tourism actions and activity from a local level, as well as building strong coalitions of local actors and dynamic connectivity into the regional/national level, and will not only meet the focus on the bottom-up methodology but will satisfactorily address the problematic failures.

We suggest that many of the problematic failures are due to the fact that, in general, top-down models tend to suffer from weakened impact, unfocused resourcing and a diffusion of structural energy at the point of local delivery due to the bureaucratic elements present in such models (Koščak and O'Rourke, 2018). The top-down method drips down resources from the top, but at the point of local delivery, the impact is potentially diminished.

Conclusion

In a sense, this chapter has painted a picture of where we were in the field of ethical and sustainable tourism in the closing months of 2019. We had described the problems that we and our contributing authors had identified, and we had made some outline of how we may face the challenges ahead.

However, this chapter in the Focus publication seeks to offer a bridge to that previous publication *Ethical and Responsible Tourism – Managing Sustainability in Local Tourism Destinations,* which in its conclusions had reflected on issues affecting ethical and responsible tourism and their future development within the framework of the local tourism environment. We continue to believe that in the "new reality" there should be a focus on:

- The three key themes of ethical and responsible tourism development
- The critical issues
- Understanding of the longer-term view

Clearly, our previous work presented a picture of where we were in the field of ethical and sustainable tourism at the end of 2019. We had described the problems identified and an outline of how we might face the challenges ahead. Those challenges are now significantly greater than we had imagined. The effect of the COVID-19 pandemic, develops from the perspective that the effect on tourism as a whole could not be foreseen. Therefore the following two chapters seek to show where we were and how we have been affected, before we consider where we may be going in what we describe as the "New Reality".

Questions

1   How may we describe the critical success factors in terms of promoting, developing and sustaining locally based cultural and heritage tourism?

2   What structure of marketing and promotion tools, delivery channels and mechanisms should be applied for the most optimal level of operational implementation in the case of local cultural and heritage tourism?

3   Which is the most efficient methodology to organise local stakeholders in the process of joint action and effective partnership? Are there differing options depending on the situation in the local destination, i.e. economic, cultural, political, etc.?

4   How important is the process of carrying capacity before and during the implementation of development and promotion in order to achieve optimal sustainability and the responsible management of cultural and heritage tourism in the destination?

5   What are the measures based on carrying capacity assessments which should sustain the actions and development of the economic, social and environmental aspects of cultural and heritage tourism in a destination?

References

ICRT (2007) Advances in responsible tourism, Occasional Paper No. 8. International Centre for Responsible Tourism, Leeds Metropolitan University, UK – www.icrtourism.org/wtm07.

Koščak, M. and O'Rourke, T. (2018) *Practical and Conceptual Strategies for the Re-evaluation of Local Tourism Destinations.* Harlow: Pearson UK.

UNESCO (2009) *Sustainable Tourism Development in UNESCO Designated Sites in South-Eastern Europe,* Ecological Tourism in Europe – ETE.

Further reading

Mariani, M. M., Buhalis, D., Longhi, C. and Vitouladiti, O. (2014) Managing change in tourism destinations: Key issues and current trends . *Journal of Destination Marketing and Management,* 2 (4), 269–272.

UNWTO (2007) *A Practical Guide to Tourism Destination Management.* Madrid: World Tourism Organization.

UNWTO and European Commission (2013) *Sustainable Tourism for Development Guidebook.* Madrid: World Tourism Organization.

2 A new tourism world

Evidence on the impacts of first and subsequent waves from March 2020 with robust restrictions on travel and accommodation show a devastating effect. International, regional and local travel restrictions affected national economies and a wide range of travel, tourism and hospitality activities. With many countries imposing travel bans, closing borders, or introducing quarantine periods, tourism declined precipitously over a period of weeks affecting virtually all parts of the hospitality value chain. The impact of cancelled events, closed accommodations, and shut down attractions became immediately felt in other parts of the supply chain, such as catering and laundry services. With the current conditions continuing the only certainty is continuing uncertainty and unknown future for tourism recovery.

Introduction

The novel coronavirus (COVID-19) continues to challenge the world with unprecedented global travel restrictions and stay-at-home orders causing the most severe disruption of the global economy since 1945. International travel bans possibly affect over 90% of the world population and wide-spread restrictions on public gatherings and mobility have resulted in tourism largely ceasing since March 2020.

Evidence of the impacts of robust restrictions on air travel, cruises, and accommodation shows a devastating effect. International, regional and local travel restrictions immediately affected national economies, including tourism systems, i.e. international travel, domestic tourism, day visits and segments as diverse as air transport, cruises, public transport, accommodation, cafés and restaurants, conventions, festivals, meetings, or sports events. With international air travel rapidly slowing as a result of the crisis, and many countries imposing travel bans, closing borders, or introducing quarantine periods, international

Table 2.1 COVID-19 cumulative data at 31 December 2020 – top five EU/ EEA countries

Country	COVID-19 14-day cumulative infections per 100,000	COVID-19 14-day deaths per 100,000
Lithuania	1199	11.5
Czechia	1119	15.6
Liechtenstein	1024	28.7
Slovenia	959	25.1
Sweden	785	3.3

Source: https://www.ecdc.europa.eu/en/covid-19-pandemic (2020)

Note: COVID-19 data at 31 December 2020; tourism data – authors' estimates at 31.12.20

and domestic tourism declined precipitously from the early months of 2020. Within countries, the virus affected all parts of the hospitality value chain. The impact of cancelled events, closed accommodation and shut down attractions impacted swiftly on other parts of the supply chain – e.g. catering and laundry services. Restaurant closures were slightly ameliorated by a switch to take-away/delivery sales allowing some to continue limited scale operations (Gössling et al., 2020).

Crisis and disaster – a crisis can be understood as an "unforeseen situation" (Demiroz and Kapucu, 2012, p. 93), and a disaster as an "unpredictable catastrophic change that can normally only be responded to after the event, either by deploying contingency plans already in place of through reactive response" (Prideaux et al., 2003). However, in agreement with Faulkner, we refer to a crisis as a "situation where the root cause of an event is, to some extent self-inflicted through such problems as inept management structures and practices or a failure to adapt to change", and disasters as "situations where an enterprise (or collection of enterprises in the case of a tourist destination) is confronted with sudden unpredictable catastrophic changes over which it has little control" (Faulkner, 2001, p.136). Disasters can be classified into five major categories, namely, political events, natural disasters, epidemics, financial events and man-made disasters. Each disaster has its own level-of-scale that needs careful study to provide various information, such as duration of disaster, the level of control, the extent of damage caused and the people affected (Laws et al., 2007; Huan et al., 2004). The effects of tourism disasters may be linked to developments in the economic, socio-cultural, political and environmental domains that affect demand and supply in destination countries. Economic downturns and recession, fluctuating exchange rates, loss of market confidence and withdrawals of investment funds are some effects of the disasters in the global tourism market. These definitions collectively allude to both planned approaches to these events and situations beyond the scope of original planning.

The situation with COVID-19 is unprecedented. Within the space of months, the framing of the global tourism system moved from over-tourism (e.g. Dodds and Butler, 2019; Seraphin et al., 2018) to non-tourism, vividly illustrated by blogs and newspaper articles depicting popular tourism sites in "before" and "after" photographs (Conde Nast Traveller, 2020). While some commentators already speculated on "What will travel be like after the Coronavirus", with some unrealistically optimistic perspectives already having proven wrong (Forbes, 2020), the general belief was that tourism would rebound as it had from previous crises (CNN, 2020). However, there is evidence that COVID-19 has been different and transformative for the tourism sector. Unlike other business sectors, tourism revenue is permanently lost because unsold capacity (e.g. accommodation) cannot be marketed in subsequent years, with corresponding implications for employment in the sector. The increasing spread of the coronavirus across countries prompted many governments to introduce unprecedented measures to contain the epidemic. These were priority measures imposed by a sanitary situation, which left little room for other options as health remained a primary concern. These measures led to many businesses being shut down temporarily, widespread restrictions on travel and mobility, financial market turmoil, an erosion of confidence and heightened uncertainty.

Global crises, such as disease outbreaks and pandemics, raise serious questions about the preparedness of global and regional tourism-related institutions to coordinate crisis management and recovery actions. The challenges are not simply economic. Issues of justice arise as vulnerable destinations and poorer populations are often disproportionately burdened by disease outbreaks. These communities often lack adequate resources to mitigate and recover from outbreaks. Vulnerabilities also exist with respect to their citizens abroad during disease outbreaks (Jamal and Budke, 2020).

The containment of the pandemic is the utmost priority and the tourism sector has committed to support all measures taken to curb the outbreak. The United Nations World Tourism Organization (UNWTO) has worked closely with the World Health Organization, UN Members States and the industry to ensure a coordinated and effective response. The COVID-19 outbreak has brought our world to a standstill with unparalleled and unforeseen impact on our lives, our economies, our societies and our livelihoods and there are growing risks of a global recession and a massive loss of jobs. Any assessment of the impact of this unparalleled crisis on the tourism sector is quickly surpassed by the fast-changing reality.

These estimates are one approach to quantifying the initial impact of containment measures on activity, and do not utilise the full range of data that inform the projections of economic growth. However, their message of sharp initial declines in activity across countries following shutdowns and restrictions on mobility is very similar to that emerging from business surveys, high-frequency daily indicators, and the sharp output contraction observed. Considerable uncertainty remains about the duration and magnitude of confinement measures and the extent to which they may be implemented in a similar manner across countries. Even once they begin to be eased, the extent of any subsequent recovery in output will depend on the effectiveness of the policy actions taken to support workers and companies through the downturn and the extent to which confidence returns.

Some figures and facts about COVID-19

Due to the fast-evolving nature of crisis, it is extremely challenging to estimate the impact of COVID-19 on international tourism. UNWTO released in March 2020 an updated assessment of the likely impact of COVID-19 on international tourism (UNWTO, 2020). Taking into account the unparalleled introduction of travel restrictions across the world, UNWTO expected that international tourist arrivals would be 20% to 30% down in 2020 by a year-on-year basis. However, UNWTO stressed that these numbers were based on the latest developments as the global community faced an unprecedented socio-economic challenge and should be interpreted with caution in view of the extreme uncertain nature of the ongoing crisis. In fact, it is highly possible that international tourism arrivals will have fallen by 70% in 2020 (year-on-year) which would be equivalent to a decline of around 1 bn global arrivals (authors' estimates). Such a fall of 70% would represent a fall in international tourism receipts (exports) of around USD1.1 tn (nominal) – i.e. some 70% of the USD1.5 tn generated in 2019 (see Tables 2.2 and 2.3).

Available data pointed to a double-digit decrease of 22% in international tourist arrivals in Q1 2020, with arrivals in the month of March down by 57% following the start of the lockdown in many countries, widespread travel restrictions and the shutdown of airports and national borders. This represented a loss of 67 million international

Table 2.2 COVID-19 effects on five high value EU tourism economies

Country	Income as percentage of GDP	COVID-19 14-day cumulative infections per 100,000	COVID-19 14-day deaths per 100,000
Croatia	18.4	422.7	20.0
Cyprus	13.9	717.7	4.6
Malta	12.7	277.2	6.1
Greece	8.7	84.2	7.3
Portugal	8.3	517.0	9.6
EU 27	*1.2*	*353*	*8.4*

Sources: https://www.ecdc.europa.eu/en/covid-19-pandemic (2020)

(High value = tourism income as percentage of GDP)
Note: COVID-19 data at 31 December 2020; tourism data – authors' estimates at 31 December 2020

Table 2.3 UNWTO forecasts – international tourist arrivals

Year	World millions	World percentage change y-o-y	Comment
2000	674	6.4%	
2001	674	**0.1%**	***Sept 11 attacks***
2002	694	3.0%	
2003	691	**-0.4%**	***SARS***
2004	756	9.4%	
2005	809	7.0%	
2006	855	5.7%	
2007	912	6.6%	
2008	929	1.9%	
2009	892	**-4.0%**	***Global economic crisis***
2010	952	6.7%	
2011	997	4.8%	
2012	1044	4.7%	
2013	1097	5.1%	
2014	1143	4.2%	
2015	1197	4.8%	
2016	1243	3.8%	

(Continued)

Table 2.3 (Cont.)

Year	World millions	World percentage change y-o-y	Comment
2017	1333	7.2%	
2018	1408	5.7%	
2019 (e)	1480	3.7%	
2020 (p)	518	**-65.0%**	*COVID-19*
2021 (p)	780	50.6%	

Source: https://www.unwto.org/international-tourism-and-covid-19

Note: (e) = estimate; (p) = projection by authors based on UNWTO data

arrivals in the first quarter of 2020 compared to the same period of the previous year. By regions, Asia and the Pacific, the first region to suffer the impact of COVID-19, saw a 35% decrease in arrivals in Q1 2020. The second-hardest hit was Europe with a 19% decline, followed by the Americas (-15%), Africa (-12%) and the Middle East (-11%).

Placing this in context, UNWTO noted that in 2009, on the back of the global economic crisis, international tourist arrivals declined by 4%, while the SARS outbreak led to a decline of just 0.4% in 2003. These estimates should of course be interpreted with caution in view of the magnitude, volatility and unprecedented nature of the current crisis; SARS and the 2009 global economic crisis were the existing references, but this crisis is unlike any other. UNWTO continues to monitor the impact of COVID-19 on international tourism. While it is too early to make a full assessment of the likely impact of COVID-19 on tourism, it is clear that millions of jobs within the sector are at risk of being lost. Around 80% of all tourism businesses are small-and-medium-sized enterprises (SMEs), and the sector has been leading the way in providing employment and other opportunities for women, youth and rural communities.

Evaluating the potential impacts/effects of COVID 19 on the tourism sector

Most major tour operators expressed what is now seen as misplaced confidence in travel picking-up post-September 2020. This was based on loose projections that around 30% of consumers would reschedule holidays to the period of June–September 2020, 40% would postpone to October–December 2020 whilst the rest would delay until 2021.

Nevertheless, it may be estimated that tour operators expected a major decline in revenue in 2020 with more than 50% seeing 2020 revenue decline by at least 50% year-on-year and 10% of operators even predicting a decline above 90%. Effectively tour operators have been caught between a rock and a hard place as travellers seek refunds whilst operators' money is tied-up with vendors (airlines, hotels, cruise lines etc.). Only half of operators seemed able to give full refunds to consumers given that around 25% were awaiting refunds from vendors. Probably only 39% of vendors were willing to give full refunds to tour operators. Tour operators, as with many low cost airlines, were offering travel credits or extending the time to complete refunds, despite consumer legislation in many countries requiring refunds to be made within a specific time scale.

Moving forward

It is clear that tourism, relying as it does on people spending time away from home, faces some particular challenges as the stricter methods of COVID-19 containment (lockdown, closing of state frontiers) are replaced from time-to-time with looser restrictions:

- Air, coach and train travel may be unattractive requiring proximity in a confined space with people who may be infectious
- Fears of the reintroduction of lockdowns or sudden quarantines
- Source markets will have higher levels of unemployment and likely higher levels of taxation; whilst travel aspiration may be undiminished, consumer assets may be more limited

For tourism to restart the source market and the destination will both need to have moved beyond lockdown and consumers will have to have both confidence and financial resources in order to travel, given that travel insurance tends not cover COVID-19 risks. There is evidence that domestic tourism and Visiting Friends and Relatives (VFR) will show signs of recovery before international travel does, but large countries are likely to maintain restrictions on internal travel. There will be competition between destinations, with offers to subsidise tourist arrivals, but the sustainability of this approach may be questionable (Goodwin and Brackenbury, 2020).

More and more, destinations are required to assess the impacts of the COVID-19 and future pandemics using both monetary and non-monetary metrics. Evaluation of these metrics should be in relation to other structurally engrained, institutional conditions and global factors such as climate change, that are increasing vulnerability of less affluent regions already struggling to build resilience after disease outbreaks and related social backlash (e.g., from Ebola, Zika virus and HIV/AIDS) (Novelli et al., 2018). The issues are related to not only economic and resource needs but also intangible injustices resulting from discrimination, racism, emotional responses and fear. Addressing these must be part of every crisis preparation, planning and response strategy. Lessons learned from past pandemics include:

- responsibility and care are needed for residents and local communities during the chaotic initial stage of a possible/actual pandemic;
- service providers and workers within the hospitality industry must be knowledgeable and prepared so that guests are not turned away inappropriately due to fear that they may be carrying COVID-19 or any similar or subsequent virus;
- communication channels must remain open between key tourism and hospitality stakeholders and the local and regional public health authorities as part of a proactive strategic response plan;
- additional research is needed on the indirect effects of health-related crises on vulnerable destinations, especially in developing countries; and
- emerging themes from the COVID-19 outbreak and from prior pandemics point to the benefit of incorporating approaches to justice (e.g., Jamal, 2019) to help inform research and practice.

(Jamal and Budke, 2020)

Shifting trends and transformation of tourism – is COVID 19 speeding these opportunities?

It is possible that after the end of this disaster, paradoxically – as with the aftermath of the 2007/8 financial crash – "soft" issues may probably fall to the bottom of the list of political priorities; issues such as:

- Sustainable development
- Climate change
- Human rights
- Refugees/economic or climate migrants

These are all contentious issues and climate change alone has the capacity to create famine, flooding and human dislocation on a scale far exceeding the COVID-19 pandemic. Destination managers together with travel agents and other industry stakeholders however have a big and sustainable opportunity at present – and the key to unlocking the opportunity is their speed in following the market, identifying and fulfilling its needs immediately.

These immediate needs, among others, will be:

· Value
· Real safety
· Lack of crowding
· Security

The mega travel and tourism organisations are generally inflexible in terms of their lack of being able to institute change quickly and decisively; cruise itineraries are set as are airline schedules. The same lack of flexibility applies to the large-scale Online Travel Agencies (OTAs) – their focus is market expansion and unit-cost reduction. Real change provides a serious challenge for them as it not only carries a cost for shareholders, but implies corporate disruption. This therefore appears to provide an opportunity for small-scale providers and connected destination managers in many unknown, remote, rural, peripheral, and little visited destinations to jointly convey the message of their existence and utilise the flexibilities and opportunities they are able to offer. The opportunities mentioned for Destination Management Organisations (DMOs) and local destinations are due to the following factors:

- As the general public may be wary of crowded places they are likely to shun air and cruise travel for a while yet. As these tourism methodologies are heavily focused on profit generation having to pay back loans and losses incurred during the COVID-19 period, whilst dealing with lower occupancy rates and lighter load factors, will inevitably increase prices. Locally focused sustainable travel and tourism has the opportunity to offer new opportunities.
- Differentiated and previously unknown destinations may become more saleable and there are potentially thousands of such destinations. DMOs and their partners have a strong and detailed awareness of the level of capacity in their destinations; this is due to the fact that unlike non-local operators, they have a detailed knowledge of their destination and its capacities. In this way clients may be impressed with unique knowledge and a unique offer. In the past most of mass tourism (99.9% of tourists) went to 0.01% of destinations – so a significant opportunity is represented.
- Slow travel – Robert Louis Stevenson is credited with having said that "to travel hopefully is a better thing than to arrive". Slow

travel means enjoying all aspects of moving slowly, considerately and ambitiously. Small-scale destinations are therefore ideal in making the travel experience calm, enjoyable, holistic and safe.

• Relationship with the client must be strong before, during and after the trip; travel should no longer be a branded commodity sold by an intermediary as part of mass-produced package. The imprint of the DMO, the intermediary and the locals at the destination in their knowledge and understanding are integral features in making an exciting and different journey. There are many uncrowded destinations that would value the potential to interact directly with the traveller; plenty of experienced operators ready to seek the business of like-minded tourists and many guides who prefer small numbers rather than massive groups.

However, first it is possible that when the pandemic has diminished, the vast majority of people will wish to travel even more. Secondly, big-company, boring homogeneous offers have never really inspired customers; rather it was the advertising copy that persuaded them to book vacation dreams that frequently replicated a well used over-tourism model. Thirdly, the industry has already set itself up for massive customer-demanded change. For example, let us look at TUI. Why is it that TUI, after over 50 years, is now no longer a tour operator? TUI, by far the most sustainably successful travel organisation in the world is now ... an "Experience Provider". The company is just one of the first-moving trailblazers in a completely new tourism economy. Inclusive tours and cruises are old, dull and totally out of fashion; they are on the way out, and now with the safety restrictions on social distance even more so. The Experience Economy is on the way in. And the great thing about the Experience Economy is that it is not dominated by the top 100 companies that until now have controlled the tourism industry – in other words not the World Travel and Tourism Council (WTTC). The Experience Economy is totally – and wonderfully – fragmented: at the moment there are some 200,000 micro-organisations run by people following their personal passions and creating some USD200bn of revenue.

Can countries and destinations use the COVID-19 crisis to make tourism better?

Many destinations have before COVID-19 encouraged unlimited growth; they have, in effect, been used by tourism. Has COVID-19 created space for reflection about tourism and about how countries and destinations wish to use tourism? Some have begun this reflection (Goodwin and

Brackenbury, 2020). Undoubtedly, international tourism activity has been seen as one of the main pathways COVID-19 may have been transmitted around the world; thus strict measures need to be in place for would-be travellers and countries to regain confidence. Countries and destinations that are virtually virus-free are unlikely to welcome hundreds of thousands of international visitors per week, arriving from countries where COVID-19 may remain prevalent. Not only because of increased infection rate spikes, but also due to infected tourists overwhelming domestic health facilities, it would be irresponsible to recommence international tourism under these circumstances.

Test and protect systems at the departure point for the outbound tourist (or pre-testing before departure) have already been tried and these systems have the possibility of enabling a restart, albeit at a limited level, of travel and tourism. On the basis that airline and hotel capacities return to normal levels, we would have to pose the question as to the extent which such surveillance systems may mitigate the fears of travellers and hotel guests. Undoubtedly some travellers will remain sceptical – those who are the more vulnerable and others concerned about civil liberties. For certain, more frequent travellers may regain confidence quickly and prices have the potential to normalise relatively quickly (Goodwin and Brackenbury, 2020).

Conclusion

Beyond the immediacy of the situation and the management of the short-term consequences of the crisis, we must now look ahead to the future, to the world of tomorrow, which will inevitably be different from all past societies and all past economies. We should make no mistake: tourism will be not be an exception. The tourism industry will have to reinvent and rethink a sustainable, digital and resilient tourism sector.

Any recovery plan, any public support for tourism, must be accompanied by transition, in order to embrace, as in all other sectors, environmental, digital and strategic realities. This was a necessity before this crisis and is now becoming a future exit imperative. But it is also necessary to create, together, a new world benchmark for responsible, sustainable and innovative tourism in response to the excesses of mass tourism, the reality of the ecological transition and the emergence of new platforms that are challenging the balance of the ecosystem.

There are likely to be three key components in such a strategy:

1 Tourism must promote sustainable tourism in the face of the "over-tourism" that can be observed in certain cities or regions. It will be a question of finding a balance between the preservation of tourist ecosystems and economic realities. We are well aware of the difficulty of such a change. It is not a question of preventing people from travelling, but of promoting, for example, local tourism and local tourism destinations. Such a change will also have to be accompanied by a new policies on tourist mobility and a strong commitment at local level.

2 Tourism will have to find a balance between the so-called traditional players and the major digital platforms. It is not a question of pitting one against the other, each will have to adapt, by becoming more digital/becoming more responsible in its role within the ecosystems accordingly.

3 To a certain extent those elements of tourism which relate to the richness of culture, history, ethnography and a priceless cultural diversity, together with the inherent protection of the environmental space, will have a strong strategic and socio-economic value.

In conclusion, we end with a short summary of the four most important points:

- A recovery of the tourism industry will not only depend on open borders, but also on changing travel behaviours – e.g. avoidance of mass transportation, large events, large/full restaurants.
- In addition general economic factors such as lower disposable incomes or an increased need to save money will have an effect.
- Cities are likely to face difficulties in the Meetings Incentives Conferences and Exhibitions (MICE) segment (this also reflects on mass events such as marathons, concerts, etc.).
- Mountain destinations focusing on mass tourism (i.e. skiing centres) are likely to face difficulties, especially in the initial phases of recovery.

Questions

1 Has COVID-19 created space for reflection about tourism and concerning how countries and destinations wish to use tourism?

2 Is it possible for countries and destinations to apply the effects of the COVID-19 crisis in order to make tourism better? In which way and with what measures?

3   What might be the new trends and approaches in destination management at the local level following the COVID-19 pandemic, in terms of providing sustainable and responsible destination products?
4   Is COVID-19 speeding the opportunities for shifting trends and transforming tourism – if so in which way and how soon this may occur?

References

Conde Nast Traveller (2020) Before and after: How coronavirus has emptied tourist attractions around the world. Retrieved 31 March 2020 from http s://www.cntravellerme.com/before-and-after-photos-tourist-attractions-duri ng-Coronavirus.

CNN (2020) What will travel look like after coronavirus? Retrieved 31 March 2020 from https://www.cnn.com/travel/article/coronavirus-travel-industry-cha nges/index.html.

Demiroz, F. and Kapucu, N. (2012) The role of leadership in managing emergencies and disasters. *European Journal of Economic and Political Studies*, 5(1), pp. 91–101.

Dodds, R. and Butler, R. (Eds.) (2019) *Overtourism: Issues, Realities and Solutions*. De Gruyter.

EU Commission (2020) Joint European Roadmap towards lifting COVID-19 containment measures. Retrieved 15 May 2020 from https://ec.europa.eu/info/ sites/info/files/communication_-_a_european_roadmap_to_lifting_coronavirus_ containment_measures_0.pdf.

Faulkner, B. (2001) Towards a framework for tourism disaster management. *Tourism Management*, 22(2), pp. 135–147.

Forbes (2020) What will travel be like after the coronavirus? Retrieved 31 March 2020 from https://www.forbes.com/sites/christopherelliott/2020/03/18/ what-will-travel-be-like-after-the-coronavirus/#4febdd623329.

Goodwin, H. and Brackenbury, M. (2020) Covid-19: Is this the time to press the reset button? Retrieved 6 May 2020 from https://hub.wtm.com/covi d-19-is-this-the-time-to-press-the-reset-button/.

Gössling, S., Scott, D. and Hall, C.M. (2020) Pandemics, tourism and global change: a rapid assessment of COVID-19. *Journal of Sustainable Tourism*. Retrieved 29 April 2020 from https://doi.org/10.1080/09669582.2020.1758708.

Huan, T.C., Beaman, J. and Shelby, L. (2004) No-escape natural disaster – mitigating impacts on tourism. *Annals of Tourism Research*, 31(2), pp. 255–273. doi:10.1016/j.annals.2003.10.003.

Jamal, T. (2019) *Justice and Ethics in Tourism*. Routledge: London and New York, NY.

Jamal, T. and Budke, C. (2020) Tourism in a world with pandemics: local-global responsibility and action. *Journal of Tourism Futures*, Emerald Publishing Limited, ISSN 2055–5911. doi:10.1108/JTF-02-2020-0014.

Laws, E., Prideaux, B. and Chon, K.S. (2007) *Crisis Management in Tourism*. CABI Publisher: Cambridge, MA. doi:10.1079/9781845930479.0000.

Novelli, M., Gussing Burgess, L., Jones, A. and Ritchie, B.W. (2018) 'No ebola...still doomed' – the ebola induced tourism crisis. *Annals of Tourism Research*, 70, pp. 76–87.

Prideaux, B., Laws, E. and Faulkner, B. (2003) Events in Indonesia: exploring the limits to formal tourism trends forecasting methods in complex crisis situations. *Tourism Management*, 24(4), pp. 475–487. doi:10.1016/S0261-5177(02)00115-2.

Seraphin, H., Sheeran, P. and Pilato, M. (2018) Over-tourism and the fall of Venice as a destination. *Journal of Destination Marketing and Management*, 9, pp. 374–376. https://doi.org/10.1016/j.jdmm.2018.01.011.

UNWTO (May 2020) International tourist numbers could fall 60–80% in 2020, UNWTO reports. Retrieved 15 May 2020 from https://www.unwto.org/news/covid-19-international-tourist-numbers-could-fall-60-80-in-2020.

Further reading

Demiroz, F. and Kapucu, N. (2012) The role of leadership in managing emergencies and disasters. *European Journal of Economic and Political Studies*, 5(1), pp. 91–101.

Gössling, S., Scott, D. and Hall, C.M. (2020) Pandemics, tourism and global change: a rapid assessment of COVID-19. *Journal of Sustainable Tourism*. Retrieved 29 April 2020 from https://doi.org/10.1080/09669582.2020.1758708.

Jamal, T. and Budke, C. (2020) Tourism in a world with pandemics: local-global responsibility and action. *Journal of Tourism Futures*, Emerald Publishing Limited, ISSN 2055-5911. doi:10.1108/JTF-02-2020-0014.

UNWTO (May 2020) International tourist numbers could fall 60–80% in 2020, UNWTO reports. Retrieved 15 May 2020 from https://www.unwto.org/news/covid-19-international-tourist-numbers-could-fall-60-80-in-2020.

3   The economic and financial consequences for tourism

Introduction

The purpose of this chapter is to look at the economic, financial and social effects of the COVID-19 pandemic within the concept of this book. We begin by scene-setting, examining the global impact at a macro-economic level of the pandemic. We look initially at the macro-economic impact at a global and European level, before considering how long the economic impact of the pandemic may last and what might become the global "new reality" post 2020.

Thereafter we look at the role of tourism in the major advanced economies and in the European Union. This is followed by a consideration of the economic problems of small scale tourism enterprises and their future status in economic, financial and social terms of seeking to operate in an ethical and responsible way. To an extent there is an underlying focus on the role of ethical and responsible tourism in providing sustainable support as and when the globalised system shows signs of faltering. Furthermore, this engages the role of ethical and responsible tourism in shifting consumption, demand and confidence towards sustainable tourism activity and replacing to some extent the previous excesses of mass tourism.

The economic effect

Overview

It is without doubt a fact that the most accomplished futurologist could not have imagined the full global impact in economic and financial terms of the COVID-19 virus. Indeed we are still uncertain about the likely outcome and for how long the economic and financial effects will continue even after the medical battle with the virus is fought and won. The International Monetary Fund very succinctly spells out the scale of the economic shock: "This crisis is like no other…the shock is large… output loss associated with this health emergency and related containment measures likely dwarfs the losses that triggered the global financial crisis. …there is continued severe uncertainty about the duration and intensity of the shock. …under current circumstances there is a very different role for economic policy" (IMF, 2020).

What the IMF terms the "Great Lockdown" is set to match the economic and financial effects of the Great Depression of the 1930s. We should also bear in mind that although the developed economies may be able to make a macro-economic bounce-back more rapidly, the emerging and developing economies will not be so swift due to systemic problems in their economic structures, the size of their informal economies and the potential for the medical crisis to last far longer. To an extent, we must also consider the knock-on effect on less-developed economies from the more developed economies. A particular example of that would be a decline in the market for long-haul tourism to South East Asian economies by European travellers affected by the economic downturn.

The only significant difference between the situation at the time of the Great Depression and the likely situation between 2020–2021 is that the world now has international multi-lateral lenders of last resort – the European Central Bank, the World Bank, the IMF and the regional development banks – which are able to supply liquidity and bolster economic and trade systems.

Global situation

Globally, the most important data concerns global economic output as indicated in Table 3.1.

Table 3.1 Global, EuroZone and G7 GDP (% change)

	2019e	2020p	2021p	2022p	2023p
Global	2.8	-4.2	4.2	4.1	3.8
EuroZone	1.3	-8.3	3.6	3.3	2.2
G7	1.6	-5.8	3.8	2.8	2.3

Sources: https://www.imf.org/en/Publications/WEO/Issues/2020/09/30/world-economic-outlook-october-2020; https://data.oecd.org/gdp/real-gdp-forecast.htm; https://ec.europa.eu/eurostat/documents/2995521/10662173/2-13112020-AP-EN.pdf/0ac3f053-f601-091d-ea21-db1ecaca7e8c modified by authors' estimates and projections

e = estimation; *p* = projections

The effect of COVID-19 in economic terms was significantly greater than anticipated; the IMF had predicted that global GDP would decline by 3.0% in the first three months of 2020; in fact, according to the first half year 2020 IMF data, we can see that global GDP fell by 4.2%. Indeed, it is likely (authors' estimates) that global GDP will be at 4.2% in 2021 rather than the projected "bounce-back" figure of 5.2% discussed by the IMF in June 2020.

Clearly, whilst global figures affecting output, GDP growth, and trade show positive outcomes in 2021, the fact remains that the projected losses incurred in 2020 will not be made up, albeit that a slower bounce-back has taken place. At the same time, questions will be asked as to the extent to which economic policies to cushion the impact of the economic and financial distress on people and enterprises arising from the COVID-19 pandemic will create fiscal as well as societal issues over the next decade. Furthermore, there will be an ongoing cost to maintaining vigilance regarding the re-emergence of the same or new forms of corona virus in the future. It must be realised that the recovery in output, growth and trade will require continuing policy support from governments and international organisations even though direct state interventions in the economy may be unwound. OECD Composite Leading Indicators provide a useful forward analysis of potential trends, as seen in Table 3.2.

The advanced economies

Conditions have been more difficult for the 36 OECD countries, which include the G20 grouping. The OECD has spoken about "unprecedented

Table 3.2 Composite Leading Indicators* (main global economies). (Month-on-month % change (m-o-m) for Apr/Jun/Aug 2020 and year-on-year change (y-o-y) for Aug 2020/Aug 2019)

Economic area	Change Apr 20/Apr 19	Change Oct 20/Oct 19
OECD	-6.6%	-0.4%
Euro Zone	-8.4%	-1.3%
G7	-6.8%	-0.3%

Source: https://www.oecd.org/sdd/leading-indicators/composite-leading-indicators-cli-oecd-october-2020.htm

*Note: Composite Leading Indicators (CLIs) are a modelling structure developed by OECD which anticipate turning points in the economic environment. In general some 6–9 months after the ups/downs predicted by the CLIs, the business cycle in economies or groups of economies will tend to follow the CLI predicted trend.

falls in GDP in most G20 economies in the second quarter of 2020" (OECD, 2020d). This resulted in G20 countries seeing a Q2/2020 fall of 6.9% over Q1/2020, which compares unfavourably with the previous 1.9% fall in the first quarter of 2009, at the height of the global financial crisis. For the G7 advanced economies, predictions are for a fall of 5.8% in growth for 2020, with a slow level of recovery around 3.8% in 2021 and the prospects of average annualised growth of 2.6% in the two following years.

Taken together, the affected sectors account for between 30–40% of total output in most economies. Allowing for only partial shutdowns in some sectors, and assuming a similar extent of shutdowns in all countries, the overall direct initial hit to the level of GDP is typically between 20–25% in many major advanced economies. Clearly this had a severe impact on tourism, travel, hotels and hospitality sectors.

Extending the same approach to other economies suggests that the impact effect of business closures could result in reductions of 15% or more in the level of output throughout the advanced economies and major emerging-market economies after the full implementation of confinement measures. In the median economy, output would be likely to decline by one-quarter. Variations in the impact effect across economies reflect differences in the composition of output. Many countries in which tourism is relatively important could potentially be affected more severely by shutdowns and limitations on travel. At the other extreme, countries with relatively sizeable agricultural and mining sectors, including oil production, may experience smaller initial effects from containment measures, although output will be subsequently hit by reduced global commodity demand.

Specific problems include Foreign Direct Investment (FDI) flows; the 13 countries reporting the greatest effect from COVID-19 in mid-June 2020 (USA, France, Germany, Italy, UK, Spain, Brazil, Russia, India, Peru, China, Hong Kong SAR and Singapore) represented over half of global FDI inward stock and almost two-thirds of outflows. Investment from those countries and into those countries in terms of the hotel, tourism and airline business alone is highly significant. Overall the pandemic is likely to reduce global FDI flows by around 40% in 2020–2021, significantly higher than the one third contraction in 2008–2009 during the financial crisis (authors' unpublished data). If this places the globalisation system into a situation of concern, then that concern will be increased by the erection of policy barriers to external investments. A new EU framework for the screening of foreign direct investments entered into force in 2019 to safeguard Europe's security and public order in relation to foreign direct investments into the Union (European Commission, 2019). It is clear such policy barriers have been utilised during the COVID-19 pandemic to block wealthy sovereign for commercial buyers picking off key strategic assets in the EU. This includes large airlines and hotel/travel groups.

The European Union

The effect on the EU, a major global economic bloc and both source and destination for strong levels of travel and tourism has been quite devastating. The EU Statistics Directorate-General (Eurostat, 2020) reported that in Q2 of 2020, the year-on-year level of growth declined by 14.7% in the EU27. For Q3 of 2020, Eurostat reported (Eurostat, 2020a), the summer recovery resulted in year-on-year growth declining by a much lower level of 4.3%. However the return to lockdown of major EU economies is likely to have a potent effect on lowering economic growth for 2020 as a whole. Further, we would project that in the EU27, household consumption will fall by near to 16% on an annual basis for 2020. This may indicate that the ability of households to engage in tourism consumption will be potentially weakened over the short-to-medium term.

Employment growth is a further key economic and social factor. On a quarter over quarter basis, in Q3/2020, EU27 employment growth fell by 2.7%, whilst year-on-year employment growth fell by 1.8%. To an extent these levels of decline seem relatively small, but we must be aware that most of the 27 countries engaged in comprehensive albeit expensive labour support schemes. As a result actual hours worked in the EU27 are estimated (authors' calculations) to have fallen by 12.1%. Further, we can see that specific EU member states' declines in GDP growth were above the EU average in Q3 on a year-on-year basis; for example Croatia and Greece fell by 14.6% (authors' projections), Spain by 8.7% and Portugal by 5.7% (Eurostat, 2020b). All four are countries with high levels of tourism inflows. The EU data is however overtaken by the much larger economic decline in the former EU member, the UK. In Q2 of 2020, UK GDP growth fell by 21.7% year-on-year and in Q3 fell by 9.6%.

Tourism in the major economies and the EU

Structure

If we look first at data from the Organisation for Economic Development and Co-operation (OECD) this shows that across its membership of 37 countries, tourism represents 4.4% of GDP, provides 6.9% of employment and represents 21.5% of services exports (Tourism Policy Responses, OECD, 2020a). This most potent factor is the role of tourism in services exports, as this is an important feature of the Current Account, and thus of the overall Balance of Payments. The relevant factors in the Balance of Payments deriving from tourism are:

• Inflows of foreign exchange (Credit to the Financial account)
• Spending of tourists (Credit to the Goods and Services account)

Thus loss of more than three-quarters of services exports in the case of Mexico would be critical to the overall economy, and reflect on the stability of the currency and its sovereign borrowing power. A lesser but similarly unpleasant situation would also occur in regard to Portugal and Spain.

Initial indicators for OECD members suggest (Tourism Policy Responses, OECD, 2020) that if tourism arrivals to these countries do not re-engage until July 2020, then they will lose 45% of tourism income. If tourism arrivals do not re-engage until September 2020, then they will lose 70% of tourism income. Initial data from the OECD, indicates the sensitivity of a number of economies to the collapse of tourism markets and tourism inflows. Those where tourism is well above OECD percentage of GDP and percentage of services exports, are indicated in Table 3.3.

The tourism labour market is a key feature in a number of countries. Table 3.4 indicates the role of tourism in national labour markets in some selected OECD European economies; this demonstrates the degree of sensitivity where tourism activity ceases in such a significant and dramatic way. Tables 3.5 and 3.6 indicate further data connected to these issues.

Table 3.3 Tourism as an economic feature

OECD member	Tourism as percentage of GDP (%)	Tourism as percentage of services exports (%)
Spain	11.8	52.3
Mexico	8.7	78.3
Iceland	8.6	47.7
Portugal	8.0	51.1
France	7.4	22.2

Source: https://www.oecd.org/cfe/tourism/ modified by authors' estimates and projections

Table 3.4 Tourism and the labour market

OECD member	Tourism as percentage of labour market (%)
All OECD	11.8
Iceland	15.7
Spain	13.5
Ireland	10.3
Greece	10.0
Portugal	9.8

Source: https://www.oecd.org/cfe/tourism/

Table 3.5 GDP changes in EU's largest tourist economies

EU member	Change in GDP growth Q3 2020/Q3 2019	Tourism earnings as percentage of EU27 tourism earnings	2018 tourism receipts as percentage of BoP
Spain	-8.7%	16.0%	0.7%
Croatia *p*	-14.6%	2.5%	15.6%
Hungary *p*	-4.7%	1.1%	2.7%
Greece *p*	-14.6%	4.6%	7.5%
Portugal	-5.7%	3.9%	6.0%
EU27 average	-4.3%	-	0.4%

Source: https://ec.europa.eu/eurostat/documents/2995521/10662173/2-13112020-AP-EN.pdf/0ac3f053-f601-091d-ea21-db1ecaca7e8c

Note: BoP = Balance of Payments expressed in terms of GDP. Where p = authors' projections based on Eurostat and other data projections

Effect

Importantly a significant element in the economic and social crisis following on the first wave of COVID-19 has been the spectre of unemployment, with its consequential effects on consumption, investment confidence and public finances. The International Labour Organization (ILO, 2020) reported that in Q4 of 2020, working hours in all European countries were down by 15.7% over Q4 2019. In many economies with high tourism dependency, the average figure would be considerably higher. Inevitably the labour market in tourism and hospitality is dominated by the mass tourism segment, as it comprises the

Table 3.6 Unemployment increases in EU's largest affected labour markets (based on seasonally adjusted data)

EU member	Increase in unemployment August 2020/August 2019	Tourism earnings as percentage of EU27 tourism earnings**
Spain	15.8%	13.5%
Ireland	58.8%	10.3%
Greece	11.5%	10.0%
Portugal	15.1%	9.8%
EU27 average	15.6%	9.0%
*Iceland**	*37.5%*	*15.7%*

Source: https://ec.europa.eu/eurostat/documents/2995521/10662173/2-13112020-AP-EN.pdf/and authors' predictions based on Eurostat and other data

Notes: * EEA member state (given for comparative purposes); ** non financial services labour market – projection by authors

largest number of jobs in the majority of the tourist economies of developed countries. For example, the German-based TUI Group had a global total of 71,500 employees at the end of 2019; the closure of European airspace and the cancellation of TUI holidays and cruises across the world is a major shock to the medium-to-long term job security of the company's employees.

Although many employees of tourism companies are on furlough or similar state programmes with public finance supporting their salaries, temporary, part-time and seasonally employed staff – especially in destinations – will have had their employment suspended and be reliant on unemployment or other social protection benefits. In Spain around 22% of tourism and hospitality employees are on temporary contracts, almost double the EU27 average (authors' estimates). The future appears bleak in terms of the mass tourism labour market – as the predicted restart of travel to many popular tourist destinations was impacted by a second wave of COVID infections in July and August 2020.

Migration has become a strong problem for developed economies, and in particular has affected a number of tourism resorts (e.g. Canary Islands, Greek islands) in terms of both the handling of refugees and the use of migrants as a temporary workforce in tourism and hospitality.

OECD has reported that in 2019, official labour migration (i.e. those with visas or work-permits, whether temporary or permanent) to OECD member states had risen by 13% with some 5.3m additional migrants (OECD, 2020b). However in the first six months of 2020, as borders closed and labour lay-offs began, the number of visas or work-permits issued in OECD countries fell by 46% year-on-year. The authors estimate that migrant workers outside the health sector may have suffered most from the economic impacts of the COVID-19 pandemic, as they would tend to lack job security. For example, the authors estimate that in Sweden over half of those unemployed or temporarily laid off in the tourism and hospitality sector were migrant workers.

A similar reduction pattern may be seen in regard to refugee flows on the transit from southern and eastern Mediterranean areas towards the northern Mediterranean countries. In the first three quarters of 2015 it is estimated (Koscak & O'Rourke, 2016) that 234,580 refugees travelled on the Eastern Mediterranean corridor from Turkey to Greece and neighbouring countries; this represented 71% of all flows over the Mediterranean corridors from Africa and Asia to Europe. In the same period of 2020, the number of refugees transiting the Eastern Mediterranean corridor stood at 15,333 – approximately 28% of total refugee transits towards Europe (FRONTEX, 2020). From 2016 the Greek islands, for example, had seen refugees arriving from Turkey onto beaches filled with tourists from Northern Europe. In November 2020, the refugees remaining on the Greek islands undoubtedly outnumbered the tourists unable to leave their Scandinavian, German or British homelands.

Added to this is a demonstrable lack of confidence in global tourism by European consumers who clearly fear tourism involving large numbers of participants. Increased levels of social distancing in operating destinations make vacations less attractive as the freedom to move and interact with others is restricted. Furthermore, social distancing also reduces the capacity of hotels, bars, restaurants and attractions. As a result fewer employees are required in the face-to-face areas, although some support activities (e.g. cleaning services) will be increased. This will inevitably lead to a decline in margins, as the mass tourism product (e.g. the Mediterranean beach holiday) has been constructed on the base of narrow margins utilising economies of scale in transport and accommodation.

In 2009, following the global financial crisis and the consequent economic recession, it was estimated by the UN World Travel Organization (UNWTO, 2010) that tourism arrivals globally decreased by 4%. In March 2020, UNWTO calculated a possible decline of tourist arrivals of between 20% and 30% year-on-year (www.unwto.org), thus five to six times the level of damage inflicted on tourism in the aftermath of the global financial crisis and consequent recession.

Recovery

The wider picture

In June 2020, as the first wave of COVID-19 showed signs of easing in Europe, the IMF wrote in its *World Economic Outlook Update* (IMF, 2020b) that as

> with the April 2020 WEO projections, there is a higher-than-usual degree of uncertainty ... The baseline projection rests on key assumptions about the fallout from the pandemic. In economies with declining infection rates, the slower recovery path in the updated forecast reflects persistent social distancing into the second half of 2020; ... damage to supply potential from the larger-than-anticipated hit to activity during the lockdown ... and a hit to productivity as surviving businesses ramp up necessary workplace safety and hygiene practices.
>
> (IMF, 2020b)

Uncertainty remained as 2020 drew to a close with both global, regional and national macro-economic and sectoral forecasts continuing to be amended due to a failure to adopt to the rapid and evolving changes in the COVID-19 public health environment. Much of the global economy, particularly in advanced economies, continues to be directed towards resourcing health care systems to combat the pandemic and to deal with the ancillary problems arising from large populations locked down into their homes. This also included direct economic and financial intervention to support enterprises and individual citizens from the economic shock of the Great Lockdown. There continue to be dichotomies in policy actions and policy results whether in the general economy or the specific instance of travel and tourism. The advanced economies have more resilient financial structures and better health systems; they are able to borrow funds at a lower price on international markets and in the case of the EU countries, they are able to benefit from European funds as well as national efforts.

Emerging market and developing economies have poorer financial structures, less developed health systems and the ability to only access borrowing at a relatively higher level of cost. In addition, we may also speculate that the macro-economic shocks from the pandemic will last longer and be more systemically profound in the developing economies. The ability of the advanced economies to provide the past levels of support to the developing economies when they are staggering to revive and revitalise their own economies and begin to repay the costs of the Great Lockdown is indeed questionable.

The scale and size of recovery is therefore difficult to predict. Uncertainties clearly exist in terms of the impact of the health crisis given that in some large and populous economies (e.g. Brazil, India and USA) the first stage of infection continued into the autumn of 2020, albeit hitting differing sectors of the population. In other economies (e.g. France and Spain) the second stage came fairly quickly after the initial recovery stage as lockdowns were eased. Parallel economic uncertainties exist affecting corporate sustainability, fiscal management, labour structures and the social crises arising from unemployment due to redundancies and business closures. The total cost is yet to be seen in economic, taxation and psychological terms. Against this socio-economic background there also exist global conflicts and challenges which are affecting all countries in some form or another.

Tourism recovery

Although general recovery may possibly appear in the final quarter of 2020, the European tourist economy in particular will be seriously damaged. If we look simply at the countries where tourism is shown as a percentage of non-financial business economy, these proportions are Greece (26%), Cyprus (20%), Ireland (14%), Croatia (13%) and Austria (13%). This level of income is highly important for the Current Account Balance and the overall Balance of Payments.

Across the EU in 2017, 11.7 million were employed in tourism representing 9% of EU employment in the non-financial business sector and 22% of employment in the services sector. Mass tourism has been the most impacted due to the lockdown and potentially from a very slow level of return in consumer confidence, inspired by threats of unemployment and general economic slowdown as well as the concept of what the "new reality" may imply in terms of mass tourism markets.

Information from Eurostat indicates how different the 2019 tourism world looked in the March to June period, the most closed down period since the appearance of COVID-19 (Eurostat, 2020). In March to June 2019, 32% of Europe's annual tourist accommodation nights took place, indicating the relative importance of this pre-peak season period. This grew, in terms of monthly figures, from 6% in March, to 7% in April, 8% in May to 11% in June. It accounted for EUR170 bn of income.

In regard to the period following the potential release of the Winter 2020–2021 lockdown restrictions, we can make some predictions in regard to a potential return of tourism confidence. In other words, are tourists likely to venture from their homes to actually travel and stay in destinations (domestic or foreign), assuming that they are able to afford the higher costs implied. It is likely that from late Spring to Winter 2021 tourists will seek certain conditions which are possible to include appropriate levels of hygiene (including disinfection) in accommodation, on transport and in public areas. This may also include severely reduced numbers using common facilities (toilets, lounge and security areas in airports, hotels and cruise ships) and restrictions involving beaches and bathing areas.

This would imply that mass tourism activities – including cruise ships, tourism flights and mass tourism resorts – will be unable to immediately meet the ability to restore traveller confidence, without making significant reductions in customer numbers. Embarking on issues concerning hygiene alone, implies significant costs not only with regard to lower people-flows (to meet potential social distancing requirements) but in terms of the staffing and equipment costs. This in itself implies that tourists will have to share a part of that cost burden, as it will not simply be possible to return to the old concept of packing in larger numbers of tourists to decrease fixed costs.

Looking at the online tourism channels, we also see changes there, although to a degree they have been more resilient than the airlines. The on-line travel booking sites booking.com and Expedia are both examples of the growth of self-packaged holidays as well as their use for standalone travel or hotel bookings. In the case of booking.com the share price – a view of how the investment market views the company's prospects – over the six months from 1 November 2019 to 30 April 2020 – fell by 29%. For Expedia, over the six months from 1 November 2019 to 30 April 2020 – the share price fell by 42% (authors' calculations from public market data). AirBNB saw bookings collapse as they had to offer refunds to customers; this also resulted in some difficult financial situations for AirBNB "hosts", many of whom in locations such as Barcelona, London, Paris and Venice, had invested huge amounts in creating what was in effect a short-term "buy-to-let" market. Estimations by Bloomberg (2020) suggested, for example, that in Paris there could be around 100,000 empty host apartments.

Confidence is also an important feature; we are still uncertain about how confident consumers will be to travel for vacations (assuming they can afford to take vacations in the first place) without fear of contagion? In the same way, how confident will business travellers be to resume business trips when it is quite obvious that airports, hotels, railway stations and trains will find social distancing costly or in the case of airlines impossible? Furthermore, it was becoming evident in late July 2020, that there was growing evidence of rates of re-infection following the easing of lockdowns (e.g. in Spanish holiday resorts); at the same time there was growing concern that the winter of 2020–2021 could bring a second wave of COVID-19; these concerns proved to be fully justified as the general level of COVID-19 infections climbed rapidly.

Elements of the new reality

Sustainable tourism and the concept of consumption

Consumption will generally be viewed as a component of the overall concept of economic product. If tourism is seen as part of such a component of economic generation, then the perceived view would be that encouraging tourism consumption brings economic benefit to the host destination and nation. But that consumption in some cases comes at an environmental and in parallel, a social cost. Wiedmann, Lenzen, Keyßer and colleagues have commented that for some five decades economic growth "has continuously increased resource use and pollutant emissions far more rapidly than these have been reduced through better technology". They suggest that richer countries "are responsible for most environmental impacts and are central to any future prospect of retreating to safer environmental conditions" and at a socio-economic and socio-cultural levels "incite consumption expansion" (Wiedmann et al., 2020).

We may cite cruise tourism as a prime example of the impact of excessive consumptive behaviour; tourism expenditure in this segment is generally driven by the wealthy 60+ age group who appear to have minimal environmental concerns and whose consumption of cruise vacations results in over-tourism and environmental degradation. The beneficial effect is mainly for the shareholders of the cruise companies and infrastructure companies (e.g. harbours and ports) which assist them.

The COVID-19 pandemic brings us to pause, reflect and reset. In pausing and then reflecting, we are required to consider the question of how we measure economic behaviour in regard to tourism. Perhaps it is not "how many will come" or "how much will they spend" but rather:

- Will they impact on the local environment positively or negatively?
- Will they impact on the social and cultural values of the local community positively or negatively?

- Will there be real benefit to the local population in economically sustainable terms?
- Will their presence and business generate long-term sustainable achievements?

This then brings us to consider what will improve the lives of the inhabitants of tourism destinations but be economically sustainable. This may be income from tourism which:

1 Is consistent and hopefully for the longer term not sensitive to external forces
2 Promotes and supports public spending on key socio-economic sectors
3 Engages with environmentally sustainable public and private investment development
4 Boosts sustainable employment
5 Provides training in key skills including those linked to the culture and heritage of the destination.

Domestic tourism – space to pause and reflect

In regard to the potential for shift to domestic tourism in response to the decline of foreign visitor inflows, the following tables (Table 3.7 and Table 3.8) are illustrative of the possible problems faced. These illustrate five OECD advanced economies where over 80% of tourists are domestic as well as another five OECD advanced economies where fewer than 37% of tourists are domestic.

Clearly domestic markets are not concerned about COVID-19 limitation controls on in-bound tourists, as domestic tourists have

Table 3.7 Domestic tourism – OECD's five largest domestic markets

OECD member	Ranking in OECD	Domestic tourism as percentage of total tourism (2020p)
Germany	1st	86
USA	2nd	85
Japan	3rd	84
Mexico	4th	83
UK	5th	81

Source: https://www.oecd.org/cfe/tourism/ and authors' projections from OECD and national data

Table 3.8 Domestic tourism – OECD's five smallest domestic markets

OECD member	Ranking in OECD	Domestic tourism as percentage of total tourism (2020p)
Poland	32nd	37
Hungary	=33rd	31
Portugal	=33rd	31
Slovenia	35th	29
Iceland	36th	26

Source: https://www.oecd.org/cfe/tourism/ and authors' projections from OECD and national data

been generally exempt from internal movement restrictions. Therefore we may be able to suggest that issues concerning the expansion of the domestic tourism market take-up are primarily:

1 Capacity
2 The attractiveness and scale of the domestic tourism market compared to external travel and tourism
3 The differential between spending and consumption inputs of foreign tourists as compared to domestic tourists
4 The perceived level of anxiety of a population towards potential infection concerns involved in foreign travel (e.g. airports, aircraft, hotels)
5 Policy measures to stimulate domestic tourism (e.g. citizen voucher schemes, discounting of travel and transport)

The main agent of recovery has therefore been in domestic markets – the staycation – as tourists switched from the uncertainties of quarantining and fears of travelling by air towards home vacations with a slant towards personal accommodation (apartments, lodges, glamping, camping) or small hotels/guest houses.

The financial impact on tourism at a local level

Small and micro-tourism businesses

Just as the global financial crisis created issues for governments, international regulatory agencies and national regulators in the assessment of banking information about sub-standard or poorly performing loans, the current COVID-19 driven economic crisis delivers even more issues of a complex variety (BiS, 2017). A significant issue is the fact that a number of European countries (i.e. the EU-27 + Norway, Iceland, Switzerland, Liechtenstein and the UK) have put into place loan guarantee programmes, where the state substantially or wholly gives a guarantee to commercial banks for a loan to a micro or small business. This obviously deals with the current situation, and effectively buys time for micro and small tourist businesses (which covers local ethical and responsible tourism) to have a degree of financial stability whilst they have no income but continue to have costs. At the same time, given the circumstances, a number of banks and credit institutions (under pressure from governments) are providing repayment holidays and also exercising forbearance. The latter term covers situations which require the lending institution to show a degree of "forbearance" by creating flexibility in the management of a loan; in other words failure to make repayments would not immediately initiate debt recovery or measures leading to liquidation of the enterprise (BiS, 2017).

This implies for a micro or small business a simple re-arrangement of the loan to meet the changed circumstances of the borrower – thus a loan re-arrangement which would not then require an impaired loan classification. Re-arrangement could involve:

- Extending a loan term
- Rescheduling the dates of principal or interest payments
- Granting new or additional periods of non-payment (grace period)
- Reducing the interest rate
- Capitalising arrears
- Changing an amortising loan to an interest payment only
- Releasing collateral or accepting lower levels of collateralisation

(BiS, 2017)

Clearly policy processes adopted by European governments which recognise the enormity of the current situation for micro and small tourism enterprises by urging banks to operate forbearance policies, providing state guaranteed loans or in many cases providing direct grants to support tourism, will be beneficial to tourism businesses in continuing to exist through periods when tourism and travel is almost non-existent. A critical point will however be in terms of the length of the current crisis; even if lockdown is lifted there is no likelihood of a return to travel within Europe until later in 2020. Travel within countries may be possible, but this will be little assistance to those small tourism businesses which rely heavily on foreign markets. Furthermore, lack of clarity at the time of writing about when the COVID-19 pandemic may end, and the capacity for the pandemic to revisit in late

2021, will not only restrict tourism activity even further, but in addition will increase financial incapacity and uncertainty.

In addition, micro and small businesses will often rely on larger enterprises for supplies, connectivity into travel systems or marketing/management support. A lengthy period during which their financial resources are seriously drained, may create problems in securing credit from suppliers or business partners. Finally, we also have to take into account the optimum period for which micro and small tourism businesses can continue without customers and the regular flow of income. Customer fears about health and safety is one issue (e.g. will customers travel by air or train to reach a tourist destination); the potential for further outbreaks of COVID-19 will also be concerning in that most travel policies no longer contain cover for illness, delays or cancellations caused by the virus. The other issue is that of economic confidence, as previously mentioned if the global recession creates heavy uncertainties about employment or the value of pensions and savings, consumers may be seriously concerned about embarking on travel plans. For micro and small enterprises the situation is further complication by the fact that in many cases they do not offer packaged facilities (i.e. flight + accommodation) which under EU and EEA regulations are covered by guarantees/insurance to refund costs.

The effects on ethical and responsible local tourism

At the same time, we would suggest that local ethical and sustainable tourism may have the fastest ability to recover as it is based at a lower economic level and does not depend so greatly on foreign inbound visitors. In the transition period, which many countries may experience as they shift from lockdown through partial lockdown to health controls at borders, internal travel centred on local tourism may see the first recovery. This could include health and spa tourism, heritage and cultural tourism, adventure and active tourism and farm/agro-tourism. Family owned and operated tourism establishments will have a faster ability to re-organise and re-position, if they receive the correct marketing and promotional advice.

To a certain extent, we may suggest that micro/small scale tourism enterprises operating at a local level with fewer than ten employees may have less sensitivity to the economic downturn and recessionary trends than larger scale enterprises. In general we make a supposition that SMEs as a whole have less resilience due to their lower levels of capital support (e.g. a shareholder or bondholder base) and their greater difficulties with cash flow.

Yet interestingly it may well be that micro/small scale tourism enterprises operating at a local level, which include family owned businesses may have greater resilience. [*LINK: Chapter 11/pp. 153–162*] Why?

- Lower labour costs – particularly if they are family owned/operated and thus the labour input is highly flexible
- Integration with other economic activities – in the case of agro-tourism, small scale tourism activities are a spin-off from agricultural actions. Over the short term they may be able to sustain themselves financially from the main activities and suspend the agro-tourism
- Socially distancing positive – in the case of agro-tourism, accommodation facilities may be self-standing and therefore appropriate for family groups. This type of tourism will be less infectious than tourism in hotels or guest houses
- Ability to tap into staycation market – whilst inter-European and cross border travel in general will remain speculative as border controls and international travel remains restrictive, domestic travel markets then make themselves available to fill the booking gap

Conclusions

In this chapter we have focused on the post-virus economic problems globally, with some examples from specific countries and discussion about the future role of ethical and responsible tourism. There is little doubt that however quickly the medical world can solve the situation through vaccination or therapies for COVID-19 treatment, the second wave of critical problems lies in the economic sphere. We are already in a situation where governments have massively committed themselves to state economic policy intervention at a stage only previously seen in socialist centrally planned command economies. Indeed in the UK, the cost of intervention into the state health system is far greater in current purchasing power terms than the cost of establishing the socialised health system in 1945 (authors' estimates). In addition to those medium-to-longer term future costs thereby implied (e.g. borrowing against future fiscal income), countries in the developed world are also having to face a future economic crisis over the short-to-medium term which may well eclipse the Great Depression of the 1930s.

In the specific case of tourism, the real pain will come in those economies having a heavy reliance on mass tourism activity which constitutes an important sector of the labour market and provides significant levels of earning in terms of services. The pain will derive from a consequent decline in the current account balance, a rise in unemployment and likely increases in essential social protection measures. This implies therefore a decline in national wealth and a fall in economic growth in real terms.

For the less developed economies, the situation is more critical; the pandemic is likely to last longer and to be exacerbated by over-crowded living spaces, lack of acute medical facilities and system failure in governmental management.

Also on the negative side there has been the wide-ranging discussion about whether tourism will follow the economic recovery through a V-shaped or U-shaped dimension. The V-shaped tourism recovery was considered in terms of tourism dropping to amazingly low levels until September but as the COVID-19 pandemic eased away, it would recover rapidly and return to normal within a short-term period (i.e. early 2021). The U-shaped recovery model posited the equally rapid drop to the bottom, followed by a period of staying at the bottom until early 2021 and then accelerating back to normality in late 2022.

Both models failed to take account of the occurrence of the second wave of COVID-19 and the consequent full-lockdowns and circuit-breakers that occurred in the advanced economies from October 2020. Furthermore, we now know where we have been, we know where we are now, but we don't know where we will be next month or next year until mass vaccination and widespread tourism testing becomes the norm. In fact, if anything, the tourism recovery model will resemble a strange W – down at the beginning, up a bit in July to September 2020, plunging down again and then if mass vaccination and widespread tourism testing occur, rising some way by November 2021, and improving through early 2022 but with a full recovery not occurring until 2023.

There is some light at the end of the tunnel, however, with regard to micro and small enterprises operating at a local level and offering ethical and responsible tourist products. Despite the possible financial issues such enterprises may face (e.g. bank lending and repayment problems), they do possess an inherent level of flexibility which mass tourism operators do not have. For example they are likely to be able switch from foreign markets to domestic markets whilst cross-border travel remains at low levels. In addition, the new normal, post-2020, is set to be more encouraging for ethical and responsible tourism as consumers react to the safety and hygiene-control problems of mass tourism. Furthermore, having endured one global crisis from a viral pandemic, tourists may begin to have anxieties about the next crisis, the effects on tourism of the global-warming catastrophe that build year by year.

Questions

1   Uncertainty is often seen as a major negative influence in terms of tourist management – for operators, for the hospitality sector and for customers – has COVID-19 added uncertainty on a massive scale?
2   Over the short term the effective huge losses in the mass tourism and cruise markets will have a serious effect on the economies of those countries which rely on income from this form of tourism. However, in the medium to longer term, what will be the effect on those economies?
3   During the pandemic, some locally focused and small-scale tourism has shown signs of being able to find new markets and display a degree of flexibility that larger tourism centres cannot display. Why is this?
4   A shift towards domestic vacations has emerged in a number of advanced economies during the period of the COVID-19 pandemic; how will this impact on the Balance of Trade and the Current Account of countries which previously relied on foreign visitor income flows?

References

BiS (Bank for International Settlements) (2017) *The Basel Committee on Banking Supervision Final Guidance on the Prudential Treatment of Problem Assets – Definitions of Non-performing Exposures and Forbearance.* Basel, 4 April 2017.

Bloomberg (2020) Will AirBNB become obsolete after the corona virus? Bloomberg New York, 02.04.20. Retrieved 20 July 2020 from https://www.bloomberg.com/opinion/articles/2020-04-02/will-airbnb-become-obsolete-after-the-coronavirus.

European Commission (2019) Press release by European Commission DG Trade dated 10.04.19 on FDI screening. Retrieved 20 July 2020 from http://trade.ec.europa.eu/doclib/press/index.cfm?id=2008.

European Commission (2020a) Eurostat press release dated 13.05.20. "EU Tourism – What a normal summer season looks like-before Covid-19". Retrieved 20 July 2020 from https://ec.europa.eu/eurostat/statistics-explained/index.php/Article_name.

European Commission (2020b) Eurostat press release dated 30.04.20. "GDP Flash release". Communication 74/2020. Retrieved 20 July 2020 from https://ec.europa.eu/eurostat/documents/2995521/10294708/2-30042020-BP-EN.pdf/52640 5c5-289c-30f5-068a-d907b7d663e6.

European Commission (2020c) Eurostat press release dates 18.04.20. "Current Account Flash release". Communication 59/2020. Retrieved 20 July 2020 from https://ec.europa.eu/eurostat/documents/2995521/10685588/2-08042020-BP-EN.pdf/6a32c5d7-cdda-5c67-9b31-18a6db0764d9.

FRONTEX (2020) *Annual Risk Analysis,* European Border and Coast Guard Agency, Warsaw 2020. ISBN 978-992-9471-9619-4.

International Labour Organization (2020) *ILO Monitor – COVID-19 and the world of work.* Fifth edition 30.06.20. ILO: Geneva. Retrieved 20 July 2020 from https://www.ilo.org/wcmsp5/groups/public/@dgreports/@dcomm/documents/briefingnote/wcms_749399.pdf.

International Monetary Fund (2020a) *World Economic Outlook, Chapter 1, April 2020.* IMF: Washington DC.

International Monetary Fund (2020b) *World Economic Outlook Update, June 2020.* IMF: Washington DC.

Koscak, M. and O'Rourke, T. (2016) Balkan Migration Crisis and the impact on tourism, ch. 2 in *Refugees Travelling on a Migrant Road.* University of Maribor, Faculty of Tourism.

OECD (2020a) Tourism Policy Responses to the coronavirus. OECD, Paris, 02.06.20. Retrieved 20 July 2020 from: http://www.oecd.org/coronavirus/policy-responses/tourism-policy-responses-to-the-coronavirus-covid-19-6466aa20/.

OECD (2020b) Press release by the Organisation for Economic Co-operation & Development dated 19.10.20 on COVID-19 crisis puts migration and progress on integration at risk. Retrieved 20 July 2020 from https://www.oecd.org/migration/covid-19-crisis-puts-migration-and-progress-on-integration-at-risk.htm.

OECD (2020c) Press release by the Organisation for Economic Co-operation & Development dated 08.09.20 on CLIs in OECD membership. OECD: Paris.

OECD (2020d) Press release by the Organisation for Economic Co-operation & Development dated 14.09.20 on GDP growth in the G20. Retrieved 20 July 2020 from www.oecd.org/sdd/na/g20-gdp-growth-Q2-2020.pdf.

UNWTO (2020) Press release by UN World Tourism Organization dated 20.03.20 on tourism arrivals. Retrieved 20 July 2020 from https://www.unwto.org/news/international-tourism-arrivals-could-fall-in-2020.

Wiedmann, T., Lenzen, M., Keyßer, L.T. et al. (2020) Scientists' warning on affluence. *Nature Communications* 11, 3107. Retrieved 20 July 2020 from https://doi.org/10.1038/s41467-020-16941-y.

Further reading

Economist Intelligence Unit (2020) EIU briefing note: Down but not out – Globalisation and the threat of Covid-19. EIU: London.

European Commission (2020) Eurostat news releases on GDP, Employment and Trade from 1 April 2020 to 6 November 2020. Available at: https://ec.europa.eu/eurostat/news/news-releases.

Eurostat (2020) News Release, EuroIndicators, 133/2020, 08.09.20 https://ec.europa.eu/eurostat/documents/2995521/10545471/2-08092020-AP-EN.pdf/43764613-3547-2e40-7a24-d20c30a20f64.

IMF (April and October 2020) *World Economic Outlook.* Available at: https://www.imf.org/en/Publications/WEO.

4 Some Reflections

Following from the topics and cases, with the use of appropriate key words, explained in *Ethical and Responsible Tourism: Managing Sustainability in Local Tourism Destinations*, this chapter seeks to reflect on some of the topics and case studies in that text-book in the light of the post COVID-19 environment. This included a focus on three thematic areas of ethical sustainable development which we see continuing to be appropriate in this new reality which we face – destination management aspects, environmental and social aspects and the business impacts.

Clearly the COVID-19 pandemic has been disastrous in terms of the loss of human life, the physical and mental strains placed on large numbers of populations across the globe who have been quarantined in their homes and in terms of the costs of dealing with the pandemic and supporting business and citizens through the period. Tourism has been comprehensively damaged, not only in advanced economies, but also in those poorer developing economies where tourism provides a vital source of income and employment.

Introduction

This chapter seeks to reflect on some of the conclusions reached in the previous publication [**LINK:** *Introduction and background/pp. 1–20*] and explained in Chapter 1. We understand that the world of tourism has changed dramatically and often catastrophically in the past year. By reflecting on our previous understanding of ethical and responsible tourism issues in the light of events over the last year, we can hope to lay the foundations of an understanding of how we need to re-conceptualise our approach towards a greater and fuller degree of sustainability in tourism.

This is structured in terms of observations on three key thematic areas of impact:

1   Destination management impacts 1. [**LINK:** *Chapter 13/pp. 183–185 & 188 – 189 & 191 – 192*]
2   Environmental and social impacts 1. [**LINK:** *Chapter 13/pp. 185–187 & 189 – 190 & 192*]
3   Business impacts 1. [**LINK:** *Chapter 13/pp. 187–188 & 190 – 191 & 192*]

Destination Management Impacts

Successful performance in *destination management* requires a new and practical tourism paradigm combining excellence, co-creation and co-operation, and high quality services. Development of innovative tourism products is aimed at increasing competitiveness, facilitating sustainable tourism development, and consequently increasing tourism turnover. Although the COVID-19 pandemic is far from being ended we strongly believe that many local destinations continue in failing to achieve such standards of destination management performance as would be suitable in the new post-pandemic environment. In other words, they plan to operate using what are evidently inappropriate models suited to high-density tourism activity, rather than more sustainable models. The pandemic demonstrated that many destinations continued to focus on unsustainable target groups of tourists (e.g. low-cost air and cruise-ships). As the pandemic began and evolved these destinations were almost completely empty and as at the end of 2020 remain so (e.g. Barcelona, Dubrovnik, Venice). The pandemic exaggerated the context of the unresolved problem – connected with poor destination management – that of over tourism. This was insufficiently and unsuccessfully addressed in many destinations, and given huge and devastating prominence by the pandemic, due to travel restrictions and public health bans introduced in many countries. The fundamental question for policy-makers is whether a destination will use tourism or be used by it. On the other hand our research proves that many unknown or under-visited destinations performed relatively well during the main summer season of 2020. Some of them even increased slightly the number of visitors or increased added value in selling local gastronomy and other products, whilst keeping the number of visitors steady and their sustainability standards as high as before the pandemic.

These and many other reasons call for more *responsibility* from all stakeholders involved in the tourism management of their local destinations, be it in normal or extreme circumstances, such as the COVID-19 pandemic. There is a growing sense that unchecked tourism spoils destinations whilst destroying cultures and environments. It is possibly now the time to rethink and restructure destination management and its performance and to commence with more responsible performance in the management of many local destinations. Both *responsibility* and *sustainability* remain as the main pillars for the future tourism development in local destinations all over the globe.

This all calls for absolute and effective collaborative *participatory planning*, which depends on a number of internal factors – adequate representation of interests, shared vision, goal accomplishment, good working relationships and open communication between destination stakeholders. This, in addition, requires strong leaders and administrative support, which every successful and responsible destination should seek and gain between partners and area stakeholders. In time of crisis, such as with the COVID-19 pandemic, this matters even more.

It is also evident that we need to be aware that the time where many destinations were following primarily one aim only, namely economic profit from tourism, has definitely passed. Now is the time for more responsible and also more professional action, which includes responsible tourism planning on all levels. This suggests that the *carrying capacity study* is necessary in order to identify environmentally and culturally sensitive areas and ensure that a tourism destination is sustainable. The purpose of a carrying capacity assessment is to ensure that tourists and day visitors attracted to the particular destination will not have a deleterious impact on the cultural or natural sites, that overcrowding will not result in visitor dissatisfaction, and that local people will not feel antagonistic towards their "guests." This is essential if tourism is to contribute to the conservation of cultural and natural heritage though the realisation of economic value and raising awareness of, and commitment to, the local patrimony. Local people must be consulted in the assessment of landscapes and cultural and natural heritage assets. It is essential to ensure that the local impact of increased heritage tourism is brought within the process of developing and marketing tourism products (Koščak and O'Rourke, 2020).

Local communities, through specific destination management teams should take decisions into their own hands. This is an approach that turns traditional "top-down" development policy on its head. Under *community-led development*, local people take the reins and form a local partnership that designs and implements an integrated development strategy. This is known as the bottom-up approach (Koščak and O'Rourke, 2020). Travellers seek more and more intimate encounters with real people and places without distraction – the focus is on promoting travel to less-travelled destinations while still providing a high-standard customer service. Many local destinations prove this to be the case in the present pandemic situation; these destinations survived the first wave of crisis with a relatively modest decrease in visits. We suggest that the future of tourism is in *experiential travel*, which is a form of tourism in which people focus on experiencing a country, city or particular place by connecting to its history, people and culture. Importantly, where such experiential assets are managed and controlled by local destination management teams, there is a more comprehensive understanding of the critical threshold for local tourism development. This implies a better understanding of the tipping point at which to pause and thus avoid destruction of socio-environmental potentials.

Environmental and Social Impacts

A number of examples in many destinations have shown that in the period of time the pandemic halted global tourism, the *environment* improved significantly in destinations which were suffering from over-tourism. One example is Dubrovnik, where for the first time in 30 years grass is growing between the stone surface on the main old town street of "Stradun". In addition *social responsibility* has been recognised as a growing element in this pandemic era; for example in Venice citizens expressed their view that the city belonged to them first and that all tourism development should respect their needs first. The fact is that the reactions of societies towards tourism are diverse – some reject changes; others inculcate them into their traditions; and some will abandon their cultural roots altogether. Whilst cultural change is unavoidable as a natural part of human culture, the sudden and forced changes that tourism often brings may cause the complete breakdown of a society and may consequently cause the loss of entire cultural traditions. Socio-cultural impacts of tourism are often hard to identify or to measure and are often the subject of personal value judgments. Generally tourism brings about changes in the value systems and behaviour of inhabitants and creates changes in the structure of communities, family relationships, collective traditional life styles, ceremonies and morality (Koščak and O'Rourke, 2020). In this time of crisis, populations globally are collaborating more frequently and reacting collectively against the threats for their local social and physical environment.

Primarily *local community* matters, particularly in tourism, and above all with tourism in the period of the COVID-19 pandemic. Consumers where able are tending to focus more on their own immediate environment when selecting the next destination for travel. One of the most important motives for responsible travelling is the desire to interact with others in local tourist destinations and to comprehend their cultures. Cultural exchange supports understanding between peoples and cultures, leads to the reduction of prejudices and thus contributes to the decrease of tension between societies. Ethical and responsible tourists have been increasingly looking for the original and authentic elements of a destination's culture. The pandemic may contribute to many local destination managers' decisions to re-evaluate local heritage and traditions, leading to a renaissance of indigenous cultures, cultural arts and crafts and the rejuvenation of events and festivals that were becoming forgotten due to modern developments and adaptation to developed economy lifestyles (Koščak and O'Rourke, 2020). Many are just a step away or next door to our homes and we failed to recognise them before the COVID outbreak.

We suggest that therefore *networking* remains a key element in achieving responsible and sustainable local tourism development. There is a clear message from many local tourism destinations for a need to ensure that tourism is developed in a way that is ecological, economic and socially sustainable. To achieve this, adequate management and monitoring must be established. In this time of crisis this is critically important and perhaps crucial for the survival of local tourism destinations. To provide an optimal solution good networking between many institutions and individuals is needed and it is important that different stakeholders involved in the tourism business are responsible for the implementation of multiple parts of the guiding principles. Governments, tourism businesses, local communities, NGOs, and tourists should all contribute in ensuring that tourism becomes more sustainable (Koščak and O'Rourke, 2020).

Business Impacts

The pandemic has had a strong effect on the tourism and travel business – both at an international and local level. Lockdowns have closed borders, accommodation and transport systems, and populations have been quarantined. Smaller tourism activities at a local level appear to have shown strong elements of *business adaptation* by encouraging staycationing domestic visitors. This has enhanced the ability to present ethical and responsible credentials to a new market segment and has assisted in helping to maintain *financial stability*.

In general small tourism businesses operating on an ethical and responsible basis at a local level may have found additional markets, but they continue to encounter problems in *access to finance*; these are existing problems and are now exacerbated to an extent by capital availability during the COVID-19 crisis. Large and medium-sized travel and tourism businesses have generally been able to access corporate financing through state-backed or intra-nationally funded support and guarantee schemes. There will in the medium-term be uncertainty in the "new reality" post-COVID world about access to state support, as governments deal with the cost of their economic support to populations and companies through the crisis. This may then heighten the need for small scale ethical and responsible tourism to explore models we have previously suggested – i.e. *developing community, co-operative and mutual systems to finance local actions.* Access to finance in the new post-COVID business reality may also impel an embracing of *investment through SMART technology-enabled systems of financing.*

We have been able in general to *assess significant market changes* – whilst tourism has been in an effective process of continual change for decades, those tourism activities and destinations (large or medium sized) which have relied on mass methods of transportation (low-cost flights or cruise ships) have been effectively paused or have been closed. Furthermore, whilst the pandemic conditions continued, they had little opportunity to shift to alternative markets. Tourism markets that are dependent on small groups or travel by rail or cycle over relatively short distances have had the opportunity to exploit new market segments.

The success of "localisation" and the general scaling down of whatever tourism activity is able to exist, has also re-engaged a better understanding of *community-driven development and management.* The crisis has displayed new strengths and opportunities for *community engagement linking the management and operation of local tourism destination organisations with the actual community.* This in turn improves the range and potential for sustainable business impacts in the future.

Conclusion

We would suggest that the models and conclusions we discussed in Chapter 1 have clearly been severely halted in their potential progress and development by the COVID-19 pandemic. With populations restrained by health protection structures and with important consumer fears about the future of business and the economy, tourism and travel seem to have become key victims. Seeking to embark on programmes to encourage greater ethical and responsible action in tourism would appear to be one of those pre-COVID ideas apparently difficult to implement. However in a reduced and constrained tourism environment, tourism that is ethical and responsible, and which is sustainable in terms of the ever-nearing climate change horizon becomes important. Those tourism organisations, destinations and communities that have been able to operate and maintain some level of financial viability will be not only appropriate exemplars, but also conceptual leaders in the medium to longer term.

Questions

1   In which way and with what measures should destination stake-holders involved in tourism management in their local destinations act and take responsibility for management – whether in normal or extreme circumstances, such as during the COVID pandemic?
2   What should be the key steps and ways to rethinking and restructuring destination management and its performance, and to commencing a more responsible performance in the management of many local destinations in the post COVID-19 world?
3   Should *responsibility* and *sustainability* remain as critical pillars for future tourism development at a local destination level as existed prior to COVID-19 or is it important that they be altered due to the newly emerging trends and standards created by COVID-19?

References

Koščak, M. and O'Rourke, T. (2018) *Practical and Conceptual Strategies for the Re-evaluation of Local Tourism Destinations*. Harlow: Pearson UK.
Koščak, M. and O'Rourke, T. (2020) Ethical and Responsible Tourism: Managing Sustainability in Local Tourism Destinations. Abingdon, UK & New York, USA: Routledge.

Further reading

Forbes (2020) What will travel be like after the coronavirus? https://www.forbes.com/sites/christopherelliott/2020/03/18/what-will-travel-be-like-after-the-coronavirus/#4febdd623329.
Goodwin, H. (2019) Personalised academic internships. [online] Accessed 27 May 2019 at: www.letsgointernship.com/en/blog/there-difference-between-sustainable-and-responsible-tourism.
OECD (2020) Press release by the Organisation for Economic Co-operation and Development dated 14.09.20 on GDP growth in the G20. Available at: www.oecd.org/sdd/na/g20-gdp-growth-Q2-2020.pdf.
UNESCO (2009) *Sustainable Tourism Development in UNESCO Designated Sites in South-Eastern Europe*. Ecological Tourism in Europe – ETE.

5 New dimensions in a post-COVID world

Introduction

Whilst this publication has been written as our response to the current global situation, an important element concerns the fundamental problems of the effect of mass tourism on environments and the damage created by such forms of tourism. To an extent, COVID-19 may be seen as a more pronounced change agent than sudden climate-related disasters, migrant flows or terrorism attacks. This is due to the wholesale economic slowdown that has affected countries and in particular the travel industry. In Europe, borders have re-appeared and transport systems are operating at 10% of capacity; populations are in isolation and any thoughts of tourism activities have been moved into 2021. Terrorist attacks in Egypt, Tunisia and Turkey caused a significant shift for 2–3 years towards the Northern Mediterranean; now tourism in the whole of the Mediterranean basin is closed. This chapter therefore looks at the issues and problems of mass tourism and its impacts prior to March 2020 and discusses what future scenario we may see emerging and more importantly what future we would like to envisage after the COVID-19 infection has significantly declined.

Over the past decade, those of us engaged in the development and preservation of sustainable tourism have become increasingly concerned about the effects of mass tourism on fragile environments. This has manifested itself in the effective destruction of cultural and social environments in tourism destinations such as Barcelona, Dubrovnik, Venice and not just these destinations alone. Over-tourism has resulted not only in tourism over-capacity and environmental damage but also in socio-economic danger to local populations. This results in excessively high living costs for basic accommodation driven by the ability for speculators to achieve high levels of short-term income from the rental of property to tourists. This then leads into local inhabitants having to leave historic city locations and move towards more distant suburbs, with subsequent increases in travel costs and the decimation of the character of city centres in the post-tourism hours.

Yet over the last year we have seen the sudden effect of an external event – the COVID-19 virus – which has created a situation where historic cities renowned for excessive tourism levels suddenly became ghost cities. This virus is the most recent and catastrophic phenomenon – other less catastrophic events to affect tourism have been:

- Migration flows from the Middle East through the Greek islands
- The impacts of sudden climate change in the Caribbean
- The results of terrorism attacks in tourism hot-spots in Egypt, Tunisia and Turkey

The local level – pursuing ethical and responsible tourism

It is important to understand the very sensitive nature of tourism demands and flows at a local destination management level. Mass tourism is constructed by an acceptance of the profound economic concepts of consumer demand and the ability of consumers to match optimum cost in the choices they make of tourism activities. At the same time, sustainable and ethical tourism is connected to the values of local environments whilst seeking to provide socio-economic advantage for host communities. Our task is to understand the sensitivities of those two contrasting models – mass tourism which may be clearly affected by global problems (migrationary trends, health issues and terrorism) – and sustainable local tourism which is less reliant and less susceptible to global trends and more related to the values and structures of localities.

At this moment in time, given our inability to predict the full effect of the COVID-19 virus on tourism, we can only speculate about the potential effects. At the same time, we have to begin to look at the effect of other forms of global crisis activity; climate issues are probably the most obvious. In Australia we have seen the results of both floods and fire on sensitive rural environments, many of which were also primarily tourism centres.

Without doubt, local tourism destinations (and how we manage those local destinations) are becoming increasingly important. It is clear that large tourism destinations are hugely sensitive to economic shifts – e.g. the rise and fall of currency values have a great effect on global tourism flows. Tourism consumers appear to be shifting away from certain types of mass tourism activity and towards tourism which has a significant sustainable and ecologically supportive base.

This leads us to try to evaluate how large scale tourism which has an apparent positive economic impact on destinations (e.g. cruise ships arriving in coastal destinations) may be swiftly affected by such fundamental trends as the current pandemic. Over the longer term how sustainable is this type of tourism activity if it is so sensitive to pandemics, financial uncertainty and socio-political stress? Does correctly maintained and developed local destination tourism have the capacity to respond more swiftly to socio-political and related economic trends?

Safety and security – a challenge to destination management in crisis situations

Tourism is one of the fastest growing industries and an important source of income for many countries. At the same time, the global tourism industry is exceptionally sensitive to external events, among them recession, terrorism, disease or natural disasters. For example, the World Travel and Tourism Council (WTTC, 2002) estimated that approximately three million people in the tourism industry lost their jobs following the outbreak of SARS in China, Hong Kong, Vietnam and Singapore, resulting in losses of over USD20 bn in terms of GDP. The WTTC (2002) also estimated that after the events of September 11, 2001, the USA lost USD92 bn dollars in travel and tourism, followed by Germany with a loss of USD25 bn and the UK with a loss of USD20 bn.

The effects of different crises on tourism have been extensively researched in the past, though usually the effect of a single crisis on a single destination is examined. These studies found that while the tourism economy is highly influenced by crises, tourism itself recovers rapidly (Keller and Bieger, 2010). The effects of a pandemic on the economy in general and on tourism in particular have been analysed and discussed in the literature after every pandemic (Burns et al., 2008; Kuo et al., 2009; Page et al., 2006). For example, international tourism to Asia was badly affected by SARS, but the size of the effect varied with the destination country (Kuo et al., 2008; McAleer et al., 2010).

In addition, since the 1980s the effects of terrorism on tourism have been extensively studied by means of empirical research. Arana and Leon (2008) found that the prevalence of tourist visits to several destinations in the Mediterranean and the Canary Islands was lower after September 11 and that the willingness to pay for a vacation decreased. Fielding and Shortland (2005) found that the number of tourists visiting Israel decreased in response to increased fatalities there.

Safety and security are one of the key components of travel planning for every tourist. Safety is considered as a state where risk and danger are minimal for an individual. And security is considered as an active protection from threats to provide risk free situations. The vital importance of both comes from our predisposition for fulfilling needs that Maslow (1943) described through the hierarchy of needs. The safety needs are positioned at second level just after primary biological needs. These two steps are essential for physical survival of an individual since we need to have basic nutrition, shelter and safety.

For tourists, safety and security are even more important, since tourists enter new and unknown situations on their journey. At home, risk can be more easily reduced and safety conditions more simply reachable. However, when travelling in a foreign country, circumstances are unpredictable and effected by many external factors. The safety and security elements that are of special concern for tourists are:

1   Travel safety – by air, road, rail, sea and river transport
2   Health safety – control over contagious diseases and epidemics, advice about vaccinations or other preventive measures, standards for food and accommodation safety etc. An emphasis must also be placed on safety and security standards at tourism sites (for example ski slopes) and events
3   Natural disasters and catastrophes – e.g. hurricanes, tsunamis, floods and avalanches
4   Environmental safety – danger from pollution, excess exploitation of natural resources and man-made catastrophes (e.g. nuclear disaster).

The following safety and security elements are classified as major safety threats in tourism by Pizam and Mansfeld (2006). Firstly, the major security element is the crime rate (e.g. theft, robbery, kidnapping, sexual violence, financial fraud, forgery, drugs, prostitution, corruption and bribery). The next element is violence and violation of public peace and order due to demonstrations and hooliganism. Even more disturbing is the outbreak of war and the consequent occurrence of illegal migration. Last but not least – contemporary tourism is highly affected by terrorist attacks. The research "Safety and security and the choice of a tourist destination" which was conducted in Slovenia in 2010 (Ciperle and Dobovšek, 2011) showed that when choosing a destination, the tourist primarily considers the destination itself, then issues about price, and thirdly about safety and security conditions. Only parents, travelling with children, place the safety and security factor first. But when regarding safety and security elements by themselves, the respondents described as the most hazardous, war and armed conflict, followed by terrorism, street criminality, accommodation safety, political stability of the country, crime in the country, travel safety and the potential risk of natural disasters.

Despite medical progress over the last centuries, infectious diseases still represent significant threats to modern societies. While some have been fought successfully and are only found within a few geographical areas (endemics), others have the ability to spread quickly from an initially limited outbreak, becoming epidemics or pandemics. The first and most crucial aspect of an epidemic or pandemic is, and will always remain, human suffering and the loss of lives. Nevertheless, the spread of a virus can also have important economic implications. A number of studies focusing on this aspect of the impact of epidemics and pandemics have found that the effects across the economy will be significant.

Rassy and Smith (2013) estimated in their paper the economic impact to the Mexican tourism sector of the H1N1 influenza pandemic, by examining tourist arrivals. The authors found that due to the virus, Mexico lost almost one million overseas visitors, which is estimated to have resulted in losses of around USD2.8 bn. This extended over a five-month period, mostly because of the slow return of European travellers to the country. Similarly, Heesoo et al. (2019) in a paper investigating the economic impact of the 2015 Middle East respiratory syndrome coronavirus (MERS-CoV) outbreak on the Republic of Korea's tourism-related industries, found that the relatively brief outbreak was associated with 2.1 million fewer non-citizen visitors, corresponding to around USD2.6 bn in lost tourism revenue.

A Keogh-Brown and Smith (2008) paper estimated that the impact of SARS on domestic tourism earnings saw losses reach USD3.5 bn in China and USD1.7 bn in Malaysia. In Zeng et al. (2005) note that, at the time of the outbreak, many provincial governments decided to close many natural attractions (e.g. nature reserves) temporarily, because, despite not being close to the outbreak centres, they had poor epidemic control systems. Given that the epidemic coincided with the peak tourism season, tourism businesses suffered significant losses.

Similarly, in their paper, Rosselló, Santana-Gallego and Awan (2017) estimated that in the case of malaria, dengue, yellow fever and ebola, the eradication of these diseases in affected countries in the Americas, Asia and Africa would result in an increase of around ten million additional tourists worldwide, which would translate into a rise in tourism expenditure of USD12 bn.

The effects of safety and security violation on tourism are varied, as stated in an article by Sönmez, Apostolopoulos and Tarlow (1999). Firstly, the life and health of the tourist is endangered and with this kind of experience, the tourist usually terminates their stay at a destination. Other tourists may cancel their reservation or transfer to an alternative destination. The negative media coverage diminishes the good image of a destination and tourist demand decreases. Furthermore, the destination alone may be destroyed and not only tourism, but also other economic structures may suffer recession.

Stagnation and decline in a tourism destination is usually measured in terms of tourist arrivals, but may also be reflected in tourist expenditures or industry profits, or in terms of the authenticity and sense of place of the attraction base. Tourist arrivals may decline in response to changing market interests, deteriorating infrastructure, competition, the loss or degradation of an attraction, and either human or natural disasters. Disaster events usually result in only temporary declines in tourist arrivals. For most destinations, however, any decline in arrivals is considered a crisis that must be managed, with the response reflecting the nature of the cause, along with a large amount of marketing and public relations (Ritchie 2009).

In order to commence the recovery phase as soon as possible, the destination should undertake various steps. Firstly, it should take care that the safety and security conditions are appropriately provided for; this covers intensified traffic control, security checks, less queuing, better health standards, attainable first aid, high environment standards etc. Secondly, the positive image of the destination should be restored by highlighting the competitive advantages of the destination – marketing special offers and nurturing good relationships with properly selected target markets and media. Furthermore the destination should increase cooperation with foreign governments and tourism representatives. Information about the improved safety and security conditions at the destination must be realistic and the tourism representatives should verify that directly at the site, whilst attending study tours at the destination. And lastly, the destination should offer some new tourism products that attract existing and new guests from abroad and from the home country. These suggestions are also considered by many authors, for example Sönmez, Apostolopoulos and Tarlow (1999).

Over-tourism only or mismanagement as well

Until the middle of March 2020, the term "over-tourism" was a term widely used in the tourism industry to describe the fate of many destinations on a global basis. The "over-tourism" phenomenon describes a destination that is overcrowded with tourists to such an extent that local inhabitants, and many visitors themselves, feel that there is an excess of tourists and that the quality of life experience of the destination is therefore inadequate, insufficient, and too low.

Over-tourism equates to unsustainable tourism, where the negative impacts for the destination, environment and local community outweigh any positive impacts. Such a level of unsustainability highlights the impacts on destinations of poor tourism planning and regulation, and the inevitable consequences of an unquestioning growth model in tourism. The phenomenon has produced a double effect: while this has led to a negative backlash against tourism, which in some cases is harmful to destinations, it has also increased awareness among diverse stakeholders groups about the need for more inclusive tourism decision-making, planning and governance (Koščak and O'Rourke, 2020).

Among destination managers, different solutions have been offered, such as dispersal of visitor flows. In addition both spatial and seasonal distribution of tourism flows are seen as a further strategy to cope with the problem. Technology might play an important role in finding sustainable solutions, although it is only a tool to be used and not the answer in itself. We are, for instance, already seeing the effective use of mobile data to influence visitor flows in some destinations. The opportunity of real time monitoring can ultimately assist capacity management, and the possible access to geo-location data can help planning visitor flows and improve supply chain management (Koščak and O'Rourke, 2020).

Over-tourism is clearly not a new concept; for decades we have seen overflowing beaches or over-populated tourist attractions in the summer months. However, it is evident that tourism has been increasing in the world and that in many destinations there was already a degree of excess in terms of the number of visitors. There are several reasons for this increased volume of travel; in particular it may be due to the greater affordability of air travel, the rapid growth of private accommodation and a strong increase in the supply of cruise ships. Additionally, it also relates to the relative wealth and propensity to travel of the baby-boomer generation that was born between 1945 and 1955. Barcelona, Venice and Dubrovnik have already been identified as suffering from excessive tourism growth as a result of these factors. Paris or London, although major tourism destinations, are less obviously in the over-tourism category – one reason may be that in these two mega-metropolitan conurbations attractions are dispersed. In the case of both Paris and London, another reason may be that they are not easily accessible for cruise ship visitors, which restricts the numbers arriving in the critical six-hour period we see in Barcelona, Venice and Dubrovnik. Technically speaking, we may assert that it is not tourist visitors that destroy tourism destinations but rather an uncontrolled and unplanned level of daily visitors. In some instances there are claims that excess saturation of tourists is due to the poor management of tourist flows, or due to insufficient infrastructure (transport systems, parking, catering and general visitor facilities) to accommodate more visitors than the carrying capacity would indicate. More specifically it may be that lack of infrastructure has its most disastrous effect in highly popular rural or remote tourism hot-spots, simply because of a lack of enforcement of capacity restrictions on tourism flows.

As a result of recent developments we see both destinations and policy makers giving tourism sustainability a far greater level of attention. The degree of understanding by government at all levels, the way in which they are engaged with the tourism industry, and the vision that key industry and government leaders have about sustainable tourism, will be crucial to cope with the key challenges.

What are the solutions to over-tourism?

In order to ensure sustainable tourism development and avoid problems of over-tourism, all voices from different stakeholder groups in a local destination must be heard and respected. In line with that fundamental concept, sustainable tourism development has to include a carrying capacity study. This critical feature of the sustainability process is an estimate of "the maximum number of people who can use a site without an unacceptable alteration in the physical environment and without an unacceptable decline in the quality of the experience to both visitors and residents" (Mathieson & Wall, 1982). The factors to be considered are:

• The physical impact of tourists
• The ecological impact of tourists
• The perceptions of overcrowding, and cultural and social impact on local residents

The carrying capacity study is central to meeting the objectives of sustainable tourism development, which is to ensure that the tourists and day visitors attracted to the particular destination will not have a deleterious impact on the cultural or natural sites, that overcrowding will not result in visitor dissatisfaction, and that local people will not feel antagonistic towards their "guests". This is essential if tourism is to contribute to the conservation of cultural and natural heritage through the realisation of economic value and raising awareness of, and commitment to, the local patrimony (Koščak and O'Rourke, 2020).

We should of course take into account that visitors are most often limited by the price of the tourism destination and offer. Decades ago, it was decided to make the Seychelles an expensive destination for anyone who wished to visit because it was clear that the islands were a relatively small geographic location and that with mass tourism flows they would quickly become overcrowded and destroyed. In Venice, the city government took the drastic steps of banning the construction of new hotels, the opening of fast-food restaurants and the traversing of the historic centre by massive cruise ships. Additionally they designed route directions for tourists and locals and prior to the COVID-19 pandemic were considering a tax for day visitors.

There are a number of other capacity flow practices: beaches have been closed in Thailand; in Barcelona, the construction of hotels and the licensing of accommodation were restricted; in Amsterdam the opening of souvenir shops and bicycle rental facilities were restricted; there was an increased tourist tax in Majorca; in Dubrovnik restrictions were put on the number of cruise ships and day visitors; at Machu Picchu it was the number of daily visitors that was restricted.

To summarise, in the professional literature there are two directions for better management of over-visited destinations:

- Raising prices
- Diversifying the tourist visit over a longer period, i.e. before and after the season

Another important consideration is that we should not give too little value to things and offer everything for free or at low prices, but rather at competitive prices. A tourist visit will always be where the attractions are, and if the infrastructure is insufficient, the experience for tourists and visitors will be bad and the feeling of overcrowding will quickly become evident. Thus, a solution may exist in the timely strategic thinking about what we seek from a tourist destination and how to satisfy as many stakeholders as possible, both locals, businesses and guests.

Overcompensating the strength of one group will inevitably lead to a level of dissatisfaction from other stakeholders, be it excessive profits and their non-dispersal to the local community or the inhibition of construction and supply by environmentalists. Successful long-term sustainable development of tourism requires compromise among all stakeholders! Even if that sounds difficult to achieve.

Conclusion

If we are to understand the COVID-19 message, the lesson is imperative, the changes are mandatory, and the results will be life-saving for us and for future generations. The rebuilding plans and revival of European tourism for the post-virus world will need to be upgraded, adapted and directed into a sustainable strategy against future pandemics and similar catastrophes.

Globalisation is gone – regional survival and self-reliability without devastating intrusions of pandemic spreading hordes of mass tourists will hopefully be initiated and clear thinking minds will prevail before the next pandemic arrives. Criticising the past with today's hindsight is totally unproductive and illogical if we do not propose a feasible alternative. Niche tourism, health and Balneo medicine tourism, and a more sustainable and responsible approach by all stakeholders involved in the tourism business are the complete package of total game-changing developments needed to replace the globalised mass tourism of past decades.

The time to change is now and the crystallisation of ideas for the new post-virus era should take place during this present pandemic attack to have a lasting and profound impact on our new emerging world after COVID-19.

Many tourism destinations will experience a dramatic reduction of activity in the short term, and will need to find alternative ways to support their local communities. Unfortunately, it is at times of crisis that the true value of tourism and its contribution to regional development and the livelihood of peripheral and insular areas becomes obvious. It is becoming obvious every day that it will be several months before we return to "normality" and it will take most of 2021 and beyond to recover the damage. This crisis will have major economic, political and socio-cultural impacts. It will also change best operational practices and change global strategies. In the meanwhile, there is an urgent call for resilience in the industry to ensure business continuity. Buhalis (2020) suggests that some of the key and immediate strategies should include:

1   Crisis management planning
2   Communication and openness through the use of technology and social media
3   Focusing on domestic and short-haul markets in the short-to-medium term
4   Support travellers to understand the key issues and encourage them to make informed decisions based on facts and evidence
5   Work with staff and suppliers to achieve realistic solutions that can meet the needs of all involved parties

6　Effective financial control with ongoing budgetary revision
7　Agile and real time management
8　Strategic thinking

The old Roman proverb "errare humanum est...perseverare diabolicum" ("To err is human...to persist is diabolical") cannot be more apt now than ever before, that is if we do not wish a future viral pandemic to be the end of us all, or at least to devastate the tourism industry beyond repair and render the collapse of those regions depending on this industry beyond revival.

Questions

1　What are the alternative ways to support local community destinations as a result of their having experienced a dramatic reduction in activity? What are the short- and long-term interventions that might be considered?
2　Discuss and amend the strategies for optimal operational practices and the changes in global strategies for tourism resilience during and after the COVID-19 pandemic. Are these strategies efficient and accurate in this present situation and as presented above? Are there any other strategies which may be introduced and implemented – and if so what are they?
3　Discuss and challenge the following statement: "The tourism of the past decades, the mass movement of tourist invasions, will definitely be an image of the past. From the perspective of the Corona pandemic, mass tourism was nothing else but an open wound in many national economies on the European continent permanently reinfected by the hordes of consumer tourists". In which way may the post COVID-19 tourism era address these issues? Is it a realistic scenario for the future of European tourism and local tourism destinations?

References

Arana, J.E. and Leon, C.J. (2008) The impact of terrorism on tourism demand. *Annals of Tourism Research*, 35 (2): 299–315.
Buhalis, D. (2020) Brace, Brace, Brace and resilience: The global tourism, travel, transportation, hospitality industry should prepare for a major impact from Coronavirus COVID-19, March. Retrieved 25 March 2020 from https://buhalis.blogspot.com/2020/03/brace-brace-brace-COVIT19.html 3.

Burns, A., Van der Mensbrugghe, D. and Timmer, H. (2008) *Evaluating the Economic Consequences of Avian Influenza.* Washington: World Bank.

Ciperle, M. and Dobovšek, B. (2011) Varnost in izbira turistične destinacije [Safety and security and the choice of a tourist destination]. Edited by Mekinc, J. and Dobovšek, B., *Varnost v turizmu* (pp. 63–89). Koper: Univerzitetna založba Annales.

Fielding, D. and Shortland, A. (2005) Are Americans more gung-ho than Europeans? Evidence from tourism in Israel during the Intifada. University of Otage Working Discussion Papers, No. 0506.

Heesoo, J., Maskery, B.A., Berro, A.D., Rotz, L.D., Yeon-Kyeng L. and Brown, C.M. (2019) Economic impact of the 2015 MERS outbreak on the Republic of Korea's tourism-related industries. *Health Security*, 17 (2), April. http://doi.org/10.1089/hs.2018.0115.

Keller, P. and Bieger, T. (2010) *Tourism Development after the Crises.* Berlin: Erich Schmidt Verlag.

Keogh-Brown, R.M. and Smith, R.D. (2008) The economic impact of SARS: How does the reality match the predictions? *Health Policy*, 88 (1): 110–120. https://doi.org/10.1016/j.healthpol.2008.03.003.

Koščak, M. and O'Rourke, T. (2020) *Ethical and Responsible Tourism: Managing Sustainabilty in Local Tourism Destinations.* Abingdon, UK and New York, USA: Routledge.

Kuo, H.I. et al. (2008) Assessing impacts of SARS and avian flu on international tourism demand to Asia. *Tourism Management*, 29: 917–928.

Kuo, H. I. et al. (2009) Estimation of the impact of avian flu on international tourism demand using Panel Data. *Tourism Economics*, 15 (3): 501–511.

Maslow, A.H. (1943) A theory of human motivation. *Psychological Review*, 50 (4): 370–396.

Mathieson, A. and Wall, G. (1982) *Tourism: Economic, Physical, and Social Impacts.* Harlow, Essex: Longman Group Limited. https://doi.org/10.1177/0047287583022001131

McAleer, M. et al. (2010) An econometric analysis of SARS and Avian Flu on international tourist arrivals to Asia. *Environmental Modeling and Software*, 25 (1): 100–106.

Page, S. et al. (2006) A case study of best practice – Visit Scotland's prepared response to an influenza pandemic. *Tourism Management*, 27 (3): 361–393.

Pizam, A. and Mansfeld, Y. (2006) Toward a theory of tourism security. *Tourism, Security and Safety: From Theory to Practice* (pp.1–28). Elsevier-Butterworth-Heinemann.

Rassy, D., Smith, R.D. (2013) The economic impact of H1N1 on Mexico's tourist and pork sectors. *Health Economics*, https://doi.org/10.1002/hec.2862.

Ritchie, B.W. (2009) *Crisis and Disaster Management for Tourism.* Clevedon, UK: Channel View Publications, 978-971-84541-84105-3.

Rosselló, J., Santana-Gallego, M. and Awan, W. (2017) Infectious disease risk and international tourism demand. *Health Policy and Planning*, May 1, 32 (4): 538–548. doi:10.1093/heapol/czw177.

Sönmez, S., Apostolopoulos, Y. and Tarlow, P. (1999) Tourism in crisis: Managing the effects of terrorism. *Journal of Travel Research*, 38 (1): 13–18.

WTTC (2002) *The Impact of Travel and Tourism on Jobs and the Economy – 2002: Extensive Summary.* World Travel and Tourism Council.

Zeng, B., Carter, R.W. and De Lacy, T. (2005) Short-term perturbations and tourism effects: the case of SARS in China. *Current Issues in Tourism*, 8 (4): 306–322, doi:10.1080/13683500508668220.

Further reading

Buhalis, D. (2020) Brace, Brace, Brace and resilience: The global tourism, travel, transportation, hospitality industry should prepare for a major impact from Coronavirus COVID-19, March. Retrieved, 25 March 2020 from https://buha lis.blogspot.com/2020/03/brace-brace-brace-COVIT19.html 3.

Delivorias, A. and Scholz, N. (2020) Economic impact of epidemics and pandemics. European Parliamentary Research Service, European Parliament. Retrieved 24 March 2020 from https://www.europarl.europa.eu/RegData/ etudes/BRIE/2020/646195/EPRS_BRI(2020)646195_EN.pdf.

Heesoo, J., Maskery, B.A., Berro, A.D., Rotz, L.D., Yeon-Kyeng L. and Brown, C.M. (2019) Economic impact of the 2015 MERS outbreak on the Republic of Korea's tourism-related industries. *Health Security*, 17 (2), April. http://doi.org/10.1089/hs.2018.0115.

Koščak, M. and O'Rourke, T. (2017) The Balkan migration crisis and its impact on tourism. *ABET Review, Juiz de Fora*, 7 (2): 20–31.

Pueyo, T. (2020) Coronavirus: the hammer and the dance. Retrieved 23 March 2020 from https://medium.com/@tomaspueyo/coronavirus-the-hammer-a nd-the-dance-be9337092b56.

Sönmez, S.F. (1998) Tourism, terrorism, and political instability. *Annals of Tourism Research*, 25 (2): 416–456.

6 The local tourism perspective

Introduction

A primary challenge of local governance, currently and in the decades to come, is to steer increasingly external, global forces towards local development so that such development achieves a shared vision with the local population. In cities, towns and villages throughout the world, the primary responsibility for this steering process rests with the institution of local government and its diverse local authorities.

Local authorities are generally responsible at the subsidiary level below national and regional governments for provision of social services, creating micro-economic infrastructure, regulating local business activity, and managing the natural environment. As a result they have many developmental instruments at their disposal. Yet in addition to these roles, perhaps the most important is that of facilitator amongst the diverse public, private and voluntary interests seeking to influence the direction of local development. Only with such a facilitator can a community of diverse interests define a shared vision and act consistently with this vision. At the local level, sustainable development is achieved by steering local development activities to simultaneously achieve three objectives:

1 Increased local social welfare
2 Greater, and more equitably distributed, local economic wealth
3 Enhanced integrity of local ecosystems (ICLEI, 1999)

Tourism is one of the many external forces influencing the direction and options for local development. The question of whether tourism can be sustainable – that is, whether it can contribute to local sustainable development – is therefore rightfully addressed. A truly legitimate and practical discussion on sustainable tourism must take place in and with the communities that are being influenced by tourist industry development. It must create accountability of the tourism industry to locally defined development visions. The true proof of "sustainable tourism" will be the sustainable development of local communities that serve as tourist destinations [**LINK:**/*Chapters 2–5*/*pp. 21–81*].

Tourism in 2020 failed to become free from the health, economic and emotional impacts generated by the COVID-19 pandemic, as spikes in infection rates affected all major tourism destinations and country/regional lockdowns continued. All indicators suggest that we are heading towards a change of paradigm in tourism, a new world that is yet to be discovered and we must understand, accept and adapt to this new context. In the social, family and personal sphere, the scale of values and perception of reality have changed. The confinement for weeks in their homes of almost half of the world's population surely promotes the need for freedom, to know and enjoy a leisure that in some cases has not been given sufficient attention, and means to discover our immediate neighbourhoods. This is where tourism brings value and the offer of possibilities as wide as each citizen determines. In recent surveys on the needs of citizens and their relationship with tourism, several "immediate" or "direct from the heart" responses stand out:

- Sense of freedom
- Open spaces
- Security
- Health
- Normality
- Hope
- Fragility and a new vision of the world
- Locality   (Hoffman, 2014)

This crisis has shown human fragility and how human planning and activity may be changed in such a short space of time that we are unable to react coherently. In a world where it seemed that the economy dominated everything, where algorithms had become the modern prophets through the knowledge of gigabytes of data, and where even the number of travellers who would arrive at a hotel could be predicted months in advance, we are severely impacted by a pandemic which makes any form of future planning useless. Despite everything, we remain connected thanks to tools such as the Internet, allowing "mobility" between the physical and virtual worlds. To a greater or lesser extent, this global problem that frightens us and forces us to take refuge in our homes is making us rethink the operation of the tourism sector in general and its relationship with the world. We strongly believe that locality becomes even more important in view of the recent COVID-19 situation as well as a shift in ecological discourse from sustainability to resilience.

Tourism generates cultural, economic and social values and it is certain that tourism activity will recover, however this depends on the professionals in the sector acquiring the capacity to align themselves with this new reality conception of the world and transmitting values of security, freedom, authenticity/locality, trust and respect for life and the planet.

Local matters: small = local is beautiful...

It is to be expected that life will strive to return to a near-normal upward curve when the pandemic crisis ends. But what does this near-normal or "new reality" imply for the travel and tourism industry? Probably not business as usual, at least not for many years with dramatically less air and cruise tourists, fewer tourists in iconic destinations, less imported food in all destinations, and appropriate carrying capacity restrictions. This could lead to far more sustainable profits, better service and far better benefits for local destinations. In all it could mean a really healthy global tourism industry with real opportunities for millions of people. It is likely that, with international travel being almost completely halted for the majority of tourism destinations, the domestic market is the only hope for at least some kind of tourism development and occupancy in the first half of 2021 before widespread vaccination takes effect.

Destinations turn local

Destination Management Organisations (DMO) are rethinking themselves to serve the needs of local populations, repositioning their websites as information portals for residents. They are updated on what restaurants are serving deliveries and takeaways, advising on loans and business support, or connecting them to charities and organisations helping the most vulnerable in their communities and in need of urgent support. Destinations that are different and previously unknown will become more marketable and there are many of these which are not crowded and better known by the local DMOs than any other agents. If local DMOs are able to conduct some research and communicate their findings, they may also be able to impress their clients with unique knowledge and offers. It should not be

forgotten that in the bulge of mass tourism in recent years probably 99.9% of tourists focused on 0.01% of destinations. This provides a remarkable opportunity for a realignment towards new and undiscovered local destinations.

Change of market place

During the traumatic period of lockdowns, the marketplace changed. A survey in the UK (European Travel Commission, 2020) indicated that individuals were developing changed attitudes – enjoying clean air and more fulfilling relationships with their habitat. This period of radical re-calibration – to pause, reflect and reset – will certainly result in a paradigm shift for the market – the millions of customers whose tastes locally based tourism seeks to fulfil. As air travel falters with increased hygiene concerns, there are opportunities for slow travel itineraries, which may make travel to a destination an exciting and romantic part of the holiday. This could also stimulate an extended length of stay. Train travel may be romantic if one chooses the right train; car travel also if it is combined with stops en-route, in some special eating spots. All of these possibilities may create a unique experience for visitors and help them to enjoy their stay holistically and safely. This sudden dramatic transformation therefore opens the door to a number of new opportunities including the following:

- Slow Tourism
- The Experience Economy
- Multi sensorial products
- Responsible local destination management

The Experience Economy

The Experience Economy is on the way in. The important feature of the Experience Economy is that it is not dominated by the top 100 companies which have until now controlled the tourism industry. The Experience Economy is totally and exceptionally fragmented: at the moment there are some 200,000 micro-organisations run by people following their personal passions and creating around USD200 bn of revenue. It offers different opportunities also for:

- Local DMOs, who seek to promote and market their local tourism destinations and providers

- Visitors who love travel and all of its opportunities and have a passion to do anything independently
- Local entrepreneurs seeking a sustainable career rather than a soul-destroying job and wish to be in control of their own destiny rather than operate within a large corporate organisation

Advantages and disadvantages of sustainable local tourism development

Tourism has been recognised as a global industry – one of the largest industry sectors in the world. As with any global industry, tourist business activities may have considerable impact on local development trends. The local impacts of the tourism industry are diverse and are often unique to the tourism sector. Tourist activities, as traditionally defined by the tourism industry, fundamentally involve the transportation and hosting of the tourism consumer in a local community, i.e., "tourist destination", where the tourist product is consumed. No other global industry structures itself in such a way that the consumer is brought to the product, rather than the product being delivered to the consumer in his or her own community. This structural difference produces unique social impacts upon the local tourist community, including the interruption of local customs and lifestyles, the spread of infectious diseases, changes in local demographics, and changes in local housing and labour markets. The primary product of tourism is not something produced by the industry. The product is often the heritage, wealth, and expected legacy of the community that serves as the tourist destination. The business activity of the tourism industry is to promote the "saleable" or appealing aspects of the community, transport non-residents into the community, manage the hospitality for and guide the activities of these visitors, and provide them with goods and services to purchase during their stay. If these business activities degrade the community's heritage and wealth, then the community suffers more directly than the consumer, who can return to his or her own community without responsibility for, or awareness of, the impacts of his/her tourist activities (ICLEI, 1999).

Tourism activities can, in particular, degrade the social and natural wealth of a community. The intrusion of large numbers of uninformed foreigners into local social systems can undermine pre-existing social relationships and values. This is particularly a problem where tourism business is centred in traditional social systems, such as isolated communities or indigenous peoples. Tourism in natural areas, euphemistically called "eco-tourism", can be a major source of degradation of

local ecological, economic and social systems. The intrusion of large numbers of foreign visitors with high-consumption and high-waste habits into natural areas, or into towns with inadequate waste management infrastructure, can produce changes to those natural areas at a rate that is far greater than imposed by local residents. These tourism-related changes are particularly deleterious when local residents rely on those natural areas for their sustenance. Resulting economic losses can encourage socially deleterious economic activities such as prostitution, crime, and migrant and child labour.

Solutions to adverse tourism impacts are to be found in the shared interest of local communities, tourism businesses, and tourism consumers to maintain the natural wealth and social heritage of the tourist destination. In the first instance, therefore, an institutional mechanism must be established, relative to each destination, to articulate and develop this sense of shared interest. To secure the legitimacy of these mechanisms, the participation of all interested local groups or interests must be guaranteed. Dialogue must take place in an open and transparent way. Experience demonstrates that if dialogue among interested parties is to have a real impact on development, it must generate accountability among these parties with regards to future investments, practices and policies. Consensus or decisions arising from dialogue must be reflected in institutional action. Only through such accountability can interests maintain a commitment to continued dialogue and a common agenda for local tourism development (ICLEI, 1999).

Experience demonstrates that this accountability should be reciprocal between individual or private interests and societal or public interests. Without such reciprocal accountability, local communities are typically forced to choose, in conflict, between private and public benefits, even if ample "win–win" development choices exist. On the one hand, accountability requires that property owners are provided with opportunity to retain the economic value of their property, either through sale or income generating activity. On the other hand, private market relationships, including property ownership, do not provide sufficient basis for social accountability related to "public goods", including ecological integrity and social heritage. Private property titles do not recognise the concepts of ecological integrity or social heritage. Excessive reliance upon private property ownership as a guide in development has in fact contributed to the deterioration of public goods.

Within such a framework of accountability, numerous instruments are available to guide local tourist development on a sustainable path. These instruments include (ICLEI, 1999):

- Heritage preservation requirements for site developments and building designs
- Programmes to exchange land and development rights from non-suitable to suitable development areas
- Private heritage and green space stewardship programmes
- Tax benefits accruing to property owners exercising sustainable development practices
- Tourism taxes and development fees to support construction and maintenance of required infrastructure

However, without true commitment to the sustainable tourism agenda, these instruments are not in themselves sufficient to prevent the steady erosion, by legally sanctioned private actions, of local natural wealth and social heritage. Therefore, support from the international, national, provincial and local levels of industry and government is essential to the success of this agenda.

The most direct way to reduce the adverse impacts of tourism-related travel is to increase opportunities for people to engage in appealing tourism activities in their own cities, regions or countries. In various cities of the world, this local tourism concept is known as "green tourism". Green tourism – in contrast to "eco-tourism", which relies on travel to distant locations – seeks to provide recreational attractions and hospitality facilities to local people within their local regions, thereby reducing tourism-related travel. While green tourism has the positive economic effect of stimulating local economic activity, it reduces the flow of foreign currency to developing nations – and any resulting economic benefits that may accrue to developing towns and cities from these revenue flows. However, short of definitive actions by the tourism industry and host countries/communities to reduce the negative environmental impacts of foreign travel and the negative social impacts of foreign tourist enclaves in developing nations, green tourism advocates will continue to build support among tourism consumers.

Tourist businesses may make a significant voluntary contribution to resolving the social costs of tourism by (ICLEI, 1999):

- Providing local resident employment and training opportunities, including in traditional trades and crafts
- Establishing purchasing guidelines that favour local goods and services procurement

- Making donations and investments in local recreational facilities, parks, cultural facilities and security operations (which also serve to improve local tourist amenities as well)
- Establishing local profit-sharing arrangements

Failing necessary voluntary measures, local governments should institute economic incentives and regulatory measures to reward best practices and prevent severe social impacts that need to be resolved at public expense. Ultimately, in the face of private establishments that demonstrate no long-term commitment to maintaining the preconditions for tourism – which include public safety, environmental quality, infrastructure maintenance, and economic justice – local authorities need to use all powers available to protect the cultural and environmental wealth that make an attractive local tourist destination.

Conclusion

The world is currently in a state of shock following the tremendous psychological and sociological impact of confinement, confinement in particular should be seen as a key factor in predicting possible future scenarios in tourism. In this new scenario full of both illusion *and* hope, the greatest enemy is uncertainty and the best way to face it is to predict the possible scenarios we will encounter, that is, which post-coronavirus scenarios we are heading towards, which will be determined more than ever by the social utility that legitimises our activity. For this we have a unique and indisputable tool: creativity. For the development of the tourism industry in a post-coronavirus scenario, the possibilities are multiple.

It should be taken into consideration that, in any case, the tourism offer should be directed towards closer markets, while at the same time offering proposals adapted to the new reality and to the tastes and priorities of the new customer base. Similarly, this crisis will show that tourists in general have acquired a certain level of social awareness, of connection with the concept of sustainability, and will appreciate tourist quality, understood not only as a simple economic value but also as satisfaction, quality in the service contracted, authenticity, and closeness to the values provided by the service. Personal safety and the safety of the society that welcomes it must be key elements in the choice of a tourist destination. Insecurity of all kinds caused by the world crisis of COVID-19 has been paradigmatic.

We anticipate several trends in tourist consumption habits that are emerging or becoming more apparent in response to the current crisis, when having in mind local tourism destinations, namely:

- Continuing travel restrictions due to changes in COVID-19 infection rate in a number of European countries will mean that many of the trips that would have been made abroad will now be made in the tourist's own country and most likely locally
- Increased environmental and social awareness: consumer concern for sustainability and social issues will continue, reinforcing the importance of environmental and social governance
- Ethics will be as important as aesthetics (beauty of destinations), as consumers will give priority to destinations that respect the environment, low-pollution transport, etc.
- Strengthening products with cultural identity and pride in showing the local and its value
- Expanding the need for participation and inclusion

(European Travel Commission, 2020)

In any case, the following points should be taken into consideration, when developing local tourism in future:

- Conditions and peculiarities of each country, region or local territory. There are no equal realities, no general solutions can be given without taking into consideration the local reality
- The local destinations must commit to aid for tourism and commercial enterprises with the aim of cushioning the effects of the crisis, considering that this aid cannot be for an indefinite period. They should have a social objective by confirming support for the continued activity of businesses and institutions and, therefore, for the jobs of workers
- The impact of the coronavirus crisis, the time of the beginning of the recovery, as well as the fact of the temporality of the destinations, are determining factors in the taking of a decision
- The participation of dynamic companies and institutions with knowledge of these territories must be urgent, since the time of paralysis derived from this crisis must be used to prepare and present tourism proposals in markets that are adequate and beneficial for these territories
- Facilitating travel and the transfer of tourists and trips by setting up systems for identifying and controlling tourists at borders, as well as asserting their safety during the trip
- Reinforcement of the advisory teams for the local tourism and commercial sector
- Promotion of courses and activities identified with local tourist destinations

(European Travel Commission, 2020)

In general, we conclude that the recovery of travel to all destinations worldwide will depend on economic factors, the speed with which travel restrictions are lifted, the health of the aviation industry, and the risk aversion of potential travellers. For instance, the pace and stability of a destination's recovery is likely to be affected by the extent to which it relies on travellers from more resilient markets, such as domestic and short-haul markets. It is also worthwhile to consider the opportunity for travel demand to increase due to changing travel patterns and preferences, such as residents choosing to take a domestic trip instead of an international holiday. A large increase in the share of domestic and short-haul travel is expected for European destinations in 2020, with elevated shares remaining into the medium term. Potential recovery by destination will vary, in part, according to reliance on these market segments and the opportunity to expand in these areas.

The resilience – likelihood of a stable and quick recovery – of travel demand is likely to be greater for destinations that rely more heavily on domestic and short-haul travellers, due to the following:

- *Lower cost of travel* for domestic and short-haul visitors, given shorter distances and more transportation options. This is especially relevant during a period when household incomes are being squeezed and large volumes of jobs are being lost.
- *Travel restrictions* are likely in 2021 to be eased first for domestic trips, and next for short-haul cross-border travel, while restrictions on longer-haul visitors may remain in place for longer. Some countries in Europe did have the potential additional benefit of being within the Schengen Zone, but during August 2020 it became obvious that this did not stop the erection of border flows between Schengen countries.
- *Uncertainty around transport* availability and costs is much greater for longer-haul travel, with a virtually exclusive reliance on air travel, while modal switch is an option for short-haul trips. This is particularly relevant given concerns about the financial viability of many airlines.
- *Other factors* may also contribute, such as a preference for travelling closer to home (as part of heightened risk aversion among travellers).

The significant disruption to the Travel and Tourism sector will lead to some changes in travel patterns. This may involve substitution of destinations due to income effects, travel restrictions or changing preferences. This can present an opportunity to target new markets and types of travellers to minimise visitor losses and speed up the recovery. This opportunity may be greatest for potential substitution from outbound to domestic travel. Further substitution from long-haul to short-haul travel may be a factor for some markets.

Questions

1   What type of tourism/tourists will destinations try to seek in the post COVID-19 situation?
2   Is there a potential for niche tourism as a future market?
3   Will mass tourism be able to operate a shortcut for fast recovery?
4   How quickly will tourists become more confident about travelling internationally into 2021–2022?
5   How likely are consumer needs and preferences to change post-pandemic?
6   How likely are travellers to return to old habits – e.g. low cost/high volume vacation experiences?

References

European Travel Commission (2020) *European Tourism Trends and Prospects – Quarterly Report* – Q2/2020, July, Brussels. Available at www.etc-corporate. org, ISSN No: 2034–9297.
Hoffman, M.L. (2014) From sustainability to resilience: Why locality matters. *Research in Urban Sociology*, 14: 341–357.doi:10.1108/S1047-004220140000014015.
ICLEI (1999) Local governments for sustainability. Retrieved 16 July 2020 from https://iclei.org/en/Home.html?gclid=EAIaIQobChMI4ejx_r-q7gIVjuR3Ch3F FwKAEAAYASAAEgJ2HfD_BwE.

Further reading

European Travel Commission (2020) *European Tourism Trends and Prospects – Quarterly Report* – Q2/2020, July, Brussels. Available at www.etc-corporate. org, ISSN No: 2034–9297.
Global Business Travel Association (GBTA) (2020) Report. Retrieved 16 July 2020 from https://www.gbta.org/.
ICLEI (1999) Tourism and Sustainable Development – Sustainable tourism: A local authority perspective, Background Paper # 3, Prepared by the International Council on Local Environmental Initiatives.
International Air Transport Association (IATA) (2020) *Market Report – Global Air Report*. Retrieved 16 July 2020 from https://www.reportlinker.com/market-report/Air-Transport/508793/Air-Transport?utm_source=adwords1&utm_

medium=cpc&utm_campaign=Transportation&utm_adgroup=Air_Transport_
Reports&gclid=EAIaIQobChMIyZqn5tTR6gIVTgCiAx2DDwRsEAAYASAA
EgJ5FvD_BwE.

Santos del Valle, A. (2020) *The Tourism Industry and the Impact of COVID
19 – Scenarios and Proposals.* Global Journey Consulting.

UNWTO (2003) *Co-operation and Partnerships in Tourism: A Global Perspective.*
World Tourism Organization. ISBN: 92-844-0601-3.

UNWTO (2019) *World Tourism Barometer, 17* (2), May. Print ISSN: 1728–1924.
Available at https://www.e-unwto.org/toc/wtobarometereng/17/2.

Wiweka, K., Setiawan, B., Wachyuni, S. S. and Adnyana, P. P. (2020) Local
perspective of community participation in lake toba as a tourism destination, *International Journal of Tourism and Hospitality Review*, 7 (1): 87–94.
https://doi.org/10.18510/ijthr.2020.7110.

World Travel and Tourism (2020) *Global Economic Impact and Trends 2020*,
WTTC, June 2020. Available at https://wttc.org/Research/Economic-Impact.

7 The Case Studies

Introduction to the Case Studies

Case Studies may be seen as a targeted methodology to bring real life situations and conditions into understanding the links between tourism theory and practice. The post-COVID world of tourism has created a situation in which the practicalities are glaringly obvious, whilst the theory which was happily embraced a year ago is now found not wholly relevant.

These Case Studies are intended to provide some practical examples and views of how ethical and responsible tourism may develop in the post-COVID new reality. Whilst they appear to have a strong European orientation, this is a reflection on the particular interests and experiences of the authors. By using areas familiar to us, we are able to make a better connection between practice and theory and to understand the environments in which the Case Studies are based.

However, we suggest that the Case Studies are diverse in nature and content, covering a range of issues and challenges including:

- The problems facing tourism in the post COVID-19 environment
- The impact of mass and over-tourism on fragile environments, and the post-COVID future
- Methodologies that may be applied to encourage ethical and responsible tourism in the longer term
- The critical value of developing sustainability in local tourism

More specifically, Case Studies 1 and 3 examine the Adriatic cities of Dubrovnik and Venice; both have been blighted by over-tourism as a result of day-trippers from nearby tourist centres and cruise tourism, where day visitor numbers vastly exceed the local populations. It is suggested that such visitors fail to make any positive impact on the local culture and economy and may indeed be categorised as voyeuristic travellers, failing to partake in the actual reality of a local tourist experience but rather watching from a distance. They are not guests of substance and value – just swiftly passing through whilst ticking a box on a global list of "attractive" destinations.

Case Study 4 considers the experiences of Malta which has over the past decade sought to move from a purely summer mass tourism centre to a more balanced vacation environment, but has nonetheless suffered from mass tourism growth and the unsightly spreading of holiday complexes. Further it also suffers from a lasting veneer of British colonialism, which tends to obliterate four millennia of multi-ethnic culture.

Case Study 2 considers the value in our new localised tourism world of wine tourism, linked to "slow food" and careful travelling with managed capacities. Wine tourism focuses on a destination which can be exemplified by the offering of wine, food and the innate all-embracing cultural heritage in the surrounding territory.

Case Study 5 examines the development of an alpine landscape park which seeks to balance the economic inputs of rural tourism in a natural environment with the potential threat provided by excessive capacity. The Case Study demonstrates how charging visitors arriving by car, bus or other motorised conveyance maintains sustainable capacity, provides an income for regeneration of traditional rural structures and an outlet for the sales of agricultural products.

Finally Case Study 6 concerns the use of voucher schemes following COVID-19 lockdowns to restore the badly affected tourism and hospitality sectors. Given that there has been a strong shift towards domestic tourism in the advanced economies, this Case Study examines the methodologies used to stimulate demand as well as their sustainability in the medium-to-longer terms. Mention is also made of voucher schemes to attract non-domestic tourists.

These Case Studies are hopefully exemplars of how we may view the new reality of the post-COVID tourism world. This is the new reality that demands that we pause–reflect–reset and look critically at the mass-tourism driven model which has overwhelmed tourism over the past 50 years. This is a model driven by the values of low-cost, low margin, high volume consumption; it has ignored the effects of such consumption-driven behaviour on fragile environments. The disaster of COVID-19 provides an opportunity to revisit and reflect; the Case Studies presented are neither illustrative nor inclusive of all tourism activity, but seek to provide a flavour of the current situation.

Case Study 1 – Dubrovnik

Introduction

In summer 2017 the tourism industry was over aken by a seemingly new crisis, dubbed "over tourism". Although this crisis appeared to take the industry by surprise, over tourism was simply a new name for an issue that has, in fact, been around for decades. Local communities took to the streets to protest that their home towns were saturated with tourists, and that tourism was now causing more harm than good. At the same time, tourists reported disappointing holiday experiences due to overcrowding, long queues, and the "Disneyfication" of destinations to the point where little of local culture or ways of life remained. Up until this point, the tourism industry, and many destinations that were obsessed with growing tourist numbers at any cost, had simply turned a blind eye to it. These same people, along with the UNWTO are now desperately casting around for solutions to over tourism. In the same way that over tourism is not new, the solutions to it are not new either; both academics and those in forward-thinking destinations have been establishing solutions for decades. The problem has not been lack of ways to solve this issue, but an unwillingness to accept there is a problem, or to commit to dealing with it.

Overview of some of the solutions

Responsible tourism

Responsible tourism is defined as tourism that creates better places to live in and to visit. As such, it places the need to improve destinations for the benefit of local people at the heart of its mission. It seeks to maximise the benefits of tourism (such as with the creation of local jobs, the conservation of natural and cultural heritage, improvements in infrastructure to benefit local people, etc.) and to minimise negative impacts (waste generation, overuse of water, damage to heritage, negative cultural impacts of visitors, etc.). If we go back to responsible tourism's forerunner, ecotourism, this movement is around 40–50 years old, and it has countless global examples of success. In some cases a more responsible approach to managing tourism can solve over tourism issues. However, in other cases we simply have a numbers problem, and a reduction in visitor numbers is essential.

Sustainability accounting

Rather than simply measuring tourist numbers without any measurement of how much money is retained locally (rather than reaching out to internationally owned hotel or cruise ship companies), or the social and environmental cost of hosting these visitors, destinations will need to develop what is known as sustainability accounting practices. Sustainability accounting represents the activities that have a direct impact on society, environment, and economic performance of an organisation. Sustainability accounting in managerial accounting contrasts with financial accounting in that managerial accounting is used for internal decision making and the creation of new policies that will have an effect on the organisation's performance at economic, ecological, and social level (known as the triple bottom line or Triple-Ps – People, Planet, Profit). Sustainability accounting is often used to generate value creation within an organisation. This measures the benefits and costs using the triple bottom line – economic, social and environmental impacts. This will reveal the net benefit, or otherwise, of tourism.

Challenges of low-cost flights

The industry must address the issue of excessively low-cost flights as a result of massive tax breaks for the aviation sector, given that a return flight from the UK to mainland Europe can cost as much as a couple of pizzas and a glass of beer. Ryanair previously stated their intention to offer free flights, although in the light of heavy costs for baggage and seating the concept of "free" is dubious. However, without doubt this has driven the rise in tourism numbers, as well as carbon emissions. The fact that aviation fuel is not subject to tax or VAT (2003 Energy Taxation Directive Article 14), and that the aviation industry has effectively been handed a massive tax subsidy, is an issue that nobody wants to talk about or do anything to address.

Carrying capacity

Carrying capacity defines the maximum number of tourists in a destination or visitor attraction that may be accommodated whilst sustaining the environment, heritage and most importantly the local population's enjoyment of their home and way of life. As a methodology it has some critics and this has led to a better approach (see the section on "Limits of acceptable change"). The criticisms of carrying capacity are in part related to the setting of an arbitrary number. Who defines this number? Is it the same all year round, for all types of tourist? Is it fixed or is it likely to change if the management of tourism improves?

Limits of acceptable change

This is a participatory approach – with local people, governments, the tourism industry, environmentalists etc. – working together to define when and how tourism begins to cause problems. These local stakeholders define a number of events they might see, or experience, indicating that tourism is becoming problematic. These are the "limits of acceptable change". When they occur, changes will need to be made. These limits may for example include – water shortages; the price of food increasing by 10% as suppliers shift focus towards catering for tourists; reduced access to markets overrun by tourists; pollution; lack of parking spaces due to higher tourist numbers and an increase in drunken behaviour. The limits of acceptable change should be reviewed periodically and do not represent a fixed set of criteria.

"De-marketing"

There are many ways to manage down tourist numbers. The first is to simply reduce the number of beds or other places to stay. Examples include refusing licenses for new hotels, reductions in Airbnb capacity or reduction in access for cruise ships. Other methods include reducing or ceasing all marketing to the over-visited hotspots, or alternatively promoting other destinations that might need (and be able to manage) increased tourist numbers. Variable pricing is a further technique that is applied to reduce demand; it may form part of a strategy to target specific types of customer groups. This might possibly sit uncomfortably with the ideal of making the world's heritage accessible to all, but without managing impacts the heritage may itself be damaged or destroyed. Thus ultimately no-one will be able to benefit. However, there are solutions to this dilemma; pricing might be differential during peak seasons, or dynamic pricing could be applied at different times of day to deter visits to museums and other attractions at peak times. Recently, many destinations have introduced a tourist tax, which serves the purpose of limiting demand but at the same time generating income that is used to manage tourism and its impacts far better.

Reducing demand in peak periods

Frequently tourists will seek to visit the same places at the same time (e.g. the main holiday periods). This results in the concentration of tourist numbers into just a few short months or even weeks, which inevitably leads to over tourism. One clearly identifiable strategy for addressing over tourism is to seek to spread tourist arrivals over wider periods of time. There are limits to how effective this can be – for example those working in the educational sector or who have school age children are restricted to educational vacation periods. However, techniques such as seasonal pricing and better promotion of shoulder seasons can reduce demand at peak times by influencing different demographic groups (e.g. the retired). Local and national governments and tourist boards have long believed that more is better. A "successful" year in tourism is generally considered to be one in which numbers have increased substantially. It is when those numbers become simply too great, with too many people concentrated into too small an area and considerably outnumbering the locals, that simple tourism becomes toxic over tourism.

Dubrovnik and the problems of over-tourism

There is no question that the success of *Game of Thrones* has drawn tourists to Dubrovnik and swelled total visitor numbers. The location filming generates 60,000 tourists a year, according to a study by Zagreb Institute of Economics (*The Dubrovnik Times*, 2017). This is not however the main reason for the over tourism this Croatian city experienced in the summer months of the last decade. A trio of culprits is in action here, the same trio that infects Venice, further up the Adriatic.

The cheap flights that have saturated Europe in recent years are one cause of over tourism, as they make it easy and affordable to reach Dubrovnik for a short break. In fact, the average stay in Dubrovnik in 2017 was less than three days. Many visitors choose to stay in accommodation booked through Airbnb – another classic over tourism player – in a peer-to-peer transaction that is often not subject to any kind of planning, permits or taxes. Alternatively, visitors are staying in guesthouses and apartments created to meet tourist demand, in buildings that once provided vital accommodation for local people.

Cruise ships are another key cause of over tourism in Dubrovnik, disgorging thousands of tourists each day. In 2017, the city received 742,000 passengers on 538 ships (DTB, 2018). These visitors typically spend very little time or money here (they may not even fit in a *Game of Thrones* tour!), yet ensure that Dubrovnik's historic streets, monuments, cafes and shops are crammed with people, creating an unpleasant experience for residents and visitors alike.

Over tourism alienates and drives out local people. In Dubrovnik at present, just around 1,600 people live in the Old Town, down from 5,000 in 1991. Homes and flats are turned over to tourist accommodation, which destroys any sense of community and inflates property prices. Everyday life has been impacted by the presence of thousands of tourists swamping the streets by day and partying by night, with local amenities and related jobs lost.

Over tourism also overloads infrastructure, damages nature, and threatens culture and heritage. In Dubrovnik, the rough limestone of the Old Town's main street, the Stradun, has been buffed by thousands of flip-flops and sandals to a marble-like finish. This becomes slippery when wet and has to be "roughed up" manually each year before the summer season by workers using sharp hammers, to create some accident-resistant texture.

The tourist influx is not the only issue; it is the amount of time tourists stay that is significant as well. Cruise ship passengers drastically swell the population of Dubrovnik in insane spikes of just a few hours, pouring onto land after breakfast and returning to the ship at 3pm for their all-inclusive dinner. While in Dubrovnik, these visitors might have a beer or an ice cream, but that is all. They contribute very little to the local economy but their presence means that those amenities that once catered for local people, or at least benefitted them financially, are gone – traditional cafes have become fast food outlets, and craft shops now sell cheap souvenir tat instead. In the context of a long cruise, Dubrovnik becomes just another stopping off point, and for every visitor delighted to be there, there's another who might be indifferent. Overheard comments in the Old Town include: "Excuse me, what country are we in?" and "What do you mean, you don't take Euros? The whole of Italy takes Euros."

In Dubrovnik, as in any other overcrowded hotspot, over tourism also threatens to kill off the goose that laid the golden egg. If travellers only experience over-sold destinations where they have a poor experience and meet jaded local people, they may potentially stop visiting altogether.

Changes planned – 2016 to 2019

In 2016, when UNESCO warned Dubrovnik that its World Heritage Status was at risk, it recommended the city restrict visitor numbers to 8,000 per day, arguing that when more than 8,000 visitors are inside the walls of the Old Town, "tourist blight" becomes inevitable. In response, in January 2017, the former mayor of Dubrovnik, Andro Vlahušić, launched a plan to limit numbers in line with UNESCO's recommendation, as well as installing 116 surveillance cameras to count the number of people entering and leaving the fortified complex.

Cruise ships docking were to be capped at two a day, with a maximum of 5,000 passengers on each ship, with staggered arrival and departure times. The intention was that this would reduce numbers by a fifth. To avoid contributing to over tourism in Dubrovnik, visitors were advised to travel smart and travel responsibly. Tourists were advised to visit outside the peak summer months of July and August if possible, and to explore hotspots such as the Old Town at the start and end of the day, when the cruise passengers were not present and the streets were quieter. The advice was to invest their time, money and consideration in Dubrovnik as well, by staying for a few nights, eating and shopping locally, and taking time to chat with residents. Additionally they were recommended to seek out the city's museums, festivals and local restaurants for an authentic experience that benefitted them, the traveller, as well as local people; to venture beyond the overcrowded Old Town to the beaches of the Lapad peninsula, which are easily reached by public bus, or to hop on a ferry and explore the islands off the coast. In this way, authorities sought to spread visitors' money further and mitigate overcrowding. Finally, visitors would be given hints to simply join a small group holiday, with a local guide who would be able to steer them to Dubrovnik's hidden corners, before travelling further into Croatia or along the coast.

The New Mayor of Dubrovnik Mato Franković, elected to the position in June 2017, took even more drastic measures just two months after being elected. The move was to protect the quality of the experience for visitors to the Croatian city. "We don't want to go with the maximum, we want to go lower than that", he said (Dalmacija danas, 2020). The new limit was intended to go further than UNESCO's recommendation of a limit of 8,000 people a day inside the hefty Medieval walls and instead place the cap at only 4,000. Franković stated that Dubrovnik needed to "reset" after a period of unchecked growth in the number of day trippers and cruise passengers that had flooded into the tiny city each day. The 2016 warning by UNESCO regarding Dubrovnik's world heritage status being at risk, with local concerns being heard more and more, underlined that the city was being blighted by the daily hordes.

The city authorities announced in January 2017 that CCTV cameras would be introduced to monitor – and, if necessary, stop – crowds passing through the city's three gates, but Franković suggested that more should be done, including cancelling cruise ship stops.

"I am not here to make people happy but to make the quality of life [in the city] better," he said. "Some of the cruise lines will disagree with what I'm saying but my main goal is to ensure quality for tourists and I cannot do it by keeping the situation as it is. We will lose money in the next two years – a million euros maybe by cutting the number of tourists – but in the future we will gain much more. We deserve to be a top quality destination" (Dalmacija danas, 2020).

The mayor is attempting to steer Dubrovnik away from the experiences of Venice and Barcelona, where tensions over numbers of tourists have led to protests and anti-visitor sentiment. He is targeting the hundreds of cruise ships that arrive at the port two miles from the Old Town. In 2016, 529 ships called at the port, bringing 799,916 passengers, up from 475 in 2015 and 463 in 2014.

In August 2016, in one day alone, 10,388 visitors bought tickets to walk Dubrovnik's ramparts, and a further record number was expected to exceed this in summer 2017. According to the eVisitor tourist check-in and check-out system, from 1 January to 24 August 2019, there were 1,004,840 arrivals in Dubrovnik, which was reached 21 days earlier than in 2018, or 14 per cent more than in 2018. There were 3,080,807 overnight stays, or six per cent more than in 2018. In addition in 2019 the year's three-millionth overnight stay in Dubrovnik was realised on 21 August, seven days earlier than in 2018, which confirms the excellent tourist season. Most of the guests come from the United Kingdom, USA, France, Germany, Spain, Croatia, Italy, Australia, Ireland and Finland.

Table CS1.1 Index of tourist visitor growth 2015–2016 by country

Country	Number	Year-on-year growth
UK	744864	+23%
Germany	235357	+10%
USA	234964	+8%
France	232725	+16%
Croatia	147229	-2%
Spain	135020	+11%
Sweden	112192	+20%
Finland	105743	+11%
Australia	97565	+9%
Italy	92061	+2%

Source: http://www.tzdubrovnik.hr/lang/en/index.html

"It was something that was not controlled and not planned", said Franković, explaining that he would cut the number of cruise ships arriving at peak time and attempt to move them away from peak times, such as the weekend. He also said he would impose limits on tour operators running day trips to the city. "Currently we have a problem with Thursday, Friday and Saturday. Those three days are complicated for us. We cannot have from 8am to 2pm, six cruise ships, then after 2pm, nothing at all. Forbidding a bigger number coming into the city at the same time will gain a quality", he said (Dalmacija danas, 2020).

Dubrovnik and COVID-19

Suddenly in March 2020 COVID-19 "resolved" all issues of over tourism, which had been the main problem for Dubrovnik over the past decade. The middle of March would normally see the early openings of the tourist season in Dubrovnik with air gateway, the southernmost airport in Croatia, awakening from its winter slumber. Every March for decades this tradition continued with the normal tourist season fully opening in Easter. However, this clearly was not a normal March, indeed it was not going to be a normal year at all. All the international airlines that flew into Dubrovnik, and would normally be preparing for their first flights of the year, were cancelled; although conditions improved with flights returning in July, the sudden blocking of returning visitors from Croatia by the four UK nations, and other countries due to a spike in COVID-19 infections from August 2020 onwards brought about a closure of flights from major markets.

With the city under quarantine due to COVID-19 until late May 2020, tourism – the lifeblood of the Croatian economy and one of Dubrovnik's main income streams – had come to a halt. All cruise ship stops in Dubrovnik were suspended by early April, and it would appear that major cruise lines are unlikely to start again until the beginning of 2021 at the earliest. The only relaxation is in some regions where vessels carrying relatively small numbers (e.g. 100 capacity vessels carrying only 50 with social distancing) will be able to operate.

On 19 March 2020, Croatia implemented a temporary ban of transit through border crossings to help prevent the spread of the virus. All the borders were closed – without travel even between Croatian cities. Mayor Mato Franković indicated that the City of Dubrovnik was monitoring the situation and looking at the effects that the spread of the virus was having on the local economy, which relies mostly on tourism. "Even during '90s warfare, Dubrovnik was not that empty. You can hear the silence and even the minor sound echoes over the cobblestone." After over tourism, he said, now there is "under tourism" (Dalmacija danas, 2020).

"Our expert services have made rough estimates of the financial structure of revenues in the first six months, and they tell us that as a unit of local self-government, we will earn less than HRK30m (EUR3.96m) in that period, but we are ready to adapt to any new situation", the mayor said. He added that this roughly estimated loss would be offset by surplus revenue generated in 2019, which would be spread over the May 2020 budget revision. "After June, we will re-examine the situation and make further decisions regarding maintaining financial stability in the public sector", the mayor announced, noting that the City must be prepared for each scenario. He also expressed the readiness of the City to support the business people through various measures. If the situation with the coronavirus extended to the main season, businesses would also be helped (Dalmacija danas, 2020).

Kristjan Staničić, director of the Croatian National Tourist Board (HTZ – Hrvatska turistička zajednica) said "The bookings for 2020's season were off to a great start in December and by mid-January we had up to twelve per cent more bookings. Since mid-January, everything has stopped. I hope that it will be better in early March. The reason for this is coronavirus, because the Germans, like everyone else, are frightened and they aren't booking because they don't know how things will progress with the virus in the future. This is not only the case for Croatia, the virus is everywhere, Greece, Turkey, Italy, all cruise ships, everyone is experiencing worse sales right now" (Dalmacija danas, 2020).

References

Data extrapolated from Republic of Croatia, Ministry of Tourism & Sport Website: https://mint.gov.hr.

The Dubrovnik Times (2017) Retrieved 15 July 2020 from https://www.thedubrovniktimes.com/news/dubrovnik/item/2751.

DTB – Dubrovnik Tourist Board (2018) Dubrovnik tourist arrivals 2011–2019. Retrieved 15 July 2020 from https://www.statista.com/statistics/886613/.

Dalmacija danas (2020) Retrieved 11 March 2020 from https://www.dalmacijadanas.hr/.

Case Study 2

Tourism in vineyard cottages

Introduction

Wine tourism [**LINK:** *Chapter 4/pp. 54–64 & Chapter 19/pp. 265–274*] may be established as a regional development tool, allowing the integration of the primary (agriculture), secondary (wine industry) and tertiary (tourism) sectors, highlighting the landscape attributes and displaying the regional "touristic terroir" singularities (Hall and Mitchell, 2002). The tourism and wine industries are increasingly identified as natural symbiotic partners (Carlsen and Charters, 2006) concerned about business and territorial sustainability. This win–win relationship must be anchored on co-operative networks, taking advantage of partnership skills and stakeholders' synergies.

As part of the tourist package, wine and food may be used to outline the image of a particular destination and represent part of its additional and diversified offering. This can be a powerful element in addressing new tourist markets as well as an opportunity for innovative and high-quality experiences within the existing products. Wine and food can also mitigate the problem of seasonality; they have the potential to extend the season of a tourist destination, encourage engagement by the local community in the processes of creating such tourist products, and enable visitors to discover destinations from new perspectives and with new features that were previously unknown (Koščak and O'Rourke, 2020).

The product of wine tourism should thus be visible and understood as a business opportunity with considerable development potential for a tourist destination. Some regions can use it to overcome economic crises; by marketing regional products, it is possible to include small producers and family businesses, which thereby generate new jobs and foster prosperity by giving added value to local products. This is a good reason for family hotels, restaurants, tourist farms, vineyard cottages etc. to include in their portfolio local and home-made products as these are both diverse and of high quality (Koščak and O'Rourke, 2020). The wine tourism product range includes not only conventional wine-growing products but increasingly the products of sustainable wine-growing and related enogastronomic services. This trend is particularly evident in countries such as Italy, France and Spain, where sustainable wine-growing is well developed.

In Slovenia, sustainable wine-growing is developing with individual wine-growers in the Styria (Štajerska) and Primorska regions. Wine-growers in the Goriška Brda region have made the most progress in this segment of tourist products and it thus comes as no surprise that Goriška Brda became the European Destination of Excellence in 2015. The region has become recognisable precisely through wine tourism. Slovenia is the country with the largest number of vineyards and wine cellars per capita. Slovenia is also a country where heritage and tradition are cherished, which is also manifested through the "Vineyard Retreats" touristic project, developed back in 1991 in

Southeast Slovenia in the region of Dolenjska. Vineyard retreats are smaller tourist facilities located in the midst of vineyards offering magnificent panoramic views. A wine cellar with stored wines lies under a modern furnished apartment.

Vineyard retreats were launched on the domestic and foreign touristic markets in 2010. Guests explore Slovenia's natural and cultural heritage, traditions and everyday life during their stay. Mostly visitors compliment the amazing nature, beautiful cultivated landscapes, warm hospitality and cleanliness. They especially enjoy the cuisine as well as the wines, some of which they often take home with them.

On worldwide touristic platforms, vineyard retreats have great feedback and are rated higher than 9 (out of 10). Although there has been a disturbing trend of abandoned vineyards in recent years, the product "Tourism in vineyard retreats" brings added value and helps to restore the exceptional cultural landscape by halting the cutting down of vineyards.

Sustainability and niche tourism matters more and more

The general global trend in the consumption segment (agriculture, cuisine and the energy sector as well as tourism) is to look for products that comply with the principles of respect for the environment, sustainability and environmental protection. This also applies to the narrow sector of wine-growing or wine-making. The experience of pioneering countries in this field shows that sustainability and integrated production are becoming increasingly important factors for the industry. Their experience also shows that the proportion of sustainable products in the portfolio of tourist products will need to expand in the future. However, a few conscientious individuals are not sufficient; instead, a well-planned and organised approach is required that will allow the concept of sustainability or sustainable production to become an integral part of the comprehensive story of a given wine and tourist destination. Sustainability is an opportunity for wine tourism; however, its economic feasibility must also be considered. If these trends turn out to be economically viable, wine-growers will no longer be able to ignore them. Vineyards and wine are components of a cultural heritage, which is connected to history and has been an essential element for the economic, social and cultural development of different wine regions. Wine culture has grown as part of the life, culture and diet of these regions since time immemorial. As a cultural symbol the importance of wine has changed over time, moving from an imperative source of nutrition to a cultural complement to food and conviviality, compatible with a healthy lifestyle. Promoting wine culture adds authenticity to its origins, and creates a product strongly linked to gastronomy, the pleasures of taste, as well as the underlying heritage (Koščak and O'Rourke, 2020).

In terms of volume, at least in Slovenia, wine tourism is a niche product. Niche tourism refers to how a specific tourism product may be tailored to meet the needs of a particular audience/market segment. Locations with specific niche products are able to establish and position themselves as niche tourism destinations. Niche tourism, through image creation, helps destinations to differentiate their tourism products and compete in an increasingly competitive and cluttered tourism environment. Through the use of the niche tourism life cycle it is clear that niche products will have different impacts, marketing challenges and contributions to destination development as they progress through it. This is also extremely important and an opportunity in the post COVID-19 environment, where tourism safety becomes paramount. Visitors will be impelled to select destinations and tourism productswhich display a low level of risk despite a larger volume of visitors.

Both natural and cultural structures are important, but attractiveness is also related to distance to markets (real and perceived). Getz (2000) has described the process and contents of wine destination development, involving attractions, services, hospitality, training, infrastructure, organisational development and a marketing plan, and we would like to add competitive value co-creation strategies. Enotourism should therefore be seen as an ecosystem, combining all different stakeholders' interests, being permeable to the external environment, influencing and being influenced, secured by mutual benefits of management networks. This complex ecosystem is sustained by three pillars, the Wine Culture, Territory/Landscape, and Tourism, combining all different stakeholders' interests inside this landscape.

When discussing the nexus between wine tourism and destination development, wine is seen to be a significant niche tourism product that acts as a key destination pull factor as it is inextricably linked to the destination and its image (Kivela and Crotts, 2006; Novelli, 2005). Wine tourism has surfaced as a growing area of special interest tourism significant to the regional tourism product and a key factor in the business strategy and development of the wineries and the supply side at the destination (Yuan et al., 2005; Carlsen and Charters, 2006). Kivela and Crotts (2006) discuss how food and wine can provide a viable alternative to destinations that cannot benefit from other more traditional forms of tourism, or substantial natural or cultural resources. Wine tourism offers rural destinations the opportunity to attract visitors who will come and spend time and financial resources liberally within their region. There exists a symbiotic relationship between wine and a tourism destination as the destination provides the wineries, and the natural and cultural backdrop that make it an ideal product for tourist consumption (Kivela and Crotts, 2006).

Tourism in vineyard cottages in Slovenia

Vineyard cottage tourism covers the Slovenian southeast regions of Dolenjska, Bela krajina and Kozjansko – Obsotelje and brings together 40 owners of vineyard cottages, who have outsourced their accommodation capacities for tourist purposes. In 2017, the Slovenian Tourist Board committed itself to be making Slovenia: "A green boutique global destination for high-end visitors seeking diverse and active experiences, peace, and personal benefits. A destination of five-star experiences" (Slovenian Tourism Strategy 2017–2021). The green, boutique and sustainable tourism certainly belongs to the product vineyard cottage tourism. The product of vineyard cottages is an inherent part of the cultural and natural heritage of the country, where owners of vineyard cottages have done their best to transform the abandoned cottages into environmentally friendly tourism accommodation packaged in an innovative tourist product. In 2011 vineyard retreat tourism was recognized as the second most innovative tourist product in Slovenia by the Slovenian Tourist Board.

In 2012 the Destination Management Organisation VisitDolenjska (2012) selected five key tourism products that were to be developed on domestic and foreign markets: Natural and Cultural Heritage, Active Holidays, Health and Relaxation, Food and Wine, Tourism in the Vineyard Cottages – as new destination tourist products. Statistics in recent years have shown that the vineyard cottage tourism offer saw an annual growth in volume and an increase of visits from different tourism markets. The number of overnight stays was as much as 60% higher in 2019 than it was in 2016, indicating that vineyard cottage tourism is on the rise and that tourists are looking for peace, nature and sustainable tourism.

Table CS2 Vineyard tourism statistics (2016–2019)

Year	Guests	Nights	Overnights
2016	802	978	3330
2017	750	974	3706
2018	840	1349	5333
2019	995	1426	5475

Source: Kompas Novo mesto, 2020

What is the tourism product about?

For Slovenia small vineyards are a typical size, which means less than 2 ha. This represents a unique cultural landscape in Europe. Slovene winegrowers can therefore be considered as "hobby wine producers". Each winegrower produces his own wine, usually several different sorts of wines. The winegrowers also gather in Winegrowers Associations which offer additional education, tastings, ratings, technical visits (local or European wide), as well as different events to their members.

Each vineyard has its own vineyard retreat, a building, where wines are processed and stored in cellars, while a nicely decorated apartment lies above. Unfortunately, most of those apartments are only used to a very small extent, perhaps only a few times each year. Every retreat also has a balcony or terrace with space to barbecue and they are usually located in stunning panoramic locations.

There are more than 30,000 vineyard retreats in Slovenia and a few years ago the idea was that these more or less forgotten assets should be revived and enriched with benefit. For this reason, the "Tourism in Vineyard Retreats" consortium was established in Dolenjska region, where owners of retreats, interested in renting them out as tourism infrastructure, are connected together. The travel agency Kompas Novo mesto was authorised to market the product for which a specially developed marketing model in the form of a so-called "dispersed hotel" was created. The agency is responsible for the complete marketing process and financial transactions, including the settlement of the tourist tax, while the retreat owners are responsible for keeping their unit well maintained.

Each guest is welcomed with homemade products (salami, bacon, bread) and an invitation to visit the cellar. During their stay, the guests also have access to this cellar, so they can consume the wine stored there. They are also informed about options for seven-day tourist programmes in the nearby region and seven-day tourist programmes around Slovenia, all prepared by Kompas Novo mesto. Many of the guests also visit neighbouring Croatia and Italy during their stay. The majority of the guests (90%) are foreign, and statistics show that each year the numbers of visitors has increased by 20–40%.

Retreat owners are closely connected with local producers; they also offer advice on natural and cultural sights of the area as well as recommending events to visit. Many of the local producers offer special discounts for vineyard retreat-guests, giving the retreats an additional role as promotors of the whole tourist region.

Visitors are especially fond of guesthouses, restaurants and farm tourism where they can taste the local cuisine. More and more retreats are upgrading their offer with saunas as well as hot- and massage tubs, which helps to extend the season, practically throughout the whole year. Adding to the touristic infrastructure of the area are many walking and cycling paths as well as possibilities for horseback riding and rowing. The number of providers is growing every year, which is also a significant encouragement for the Association members.

Positive effects of such products

Tourism in vineyard retreats has exceptionally positive effects:

preservation and restoration of traditional housing heritage, without interventions in existing space
new, unique accommodation facilities
the wine is marketed together with the facility
vineyards are preserved despite the recent trends of cutting down vine trees in the region

Although older generations considered vine growing and wine processing as a hobby, entertainment and way of life, the younger generations are unfortunately losing interest in vine growing. This often leads to vineyards being cut down after the older generation can no longer take care of them. It is a fact that most of the vineyards are in very steep areas where it is not possible to grow any other crops, and often the abandoned vineyards are overgrown by bushes and thorns. In a few decades, this can cause enormous damage to the appearance of the cultural landscape, of which local inhabitants are so proud. It is important to say that tourist arrivals enhance the beauty and tidiness of the vineyards and, even more importantly, the exceptional landscapes of Slovenia's countryside are preserved.

Marketing

Some retreats are occupied already for 3–4 months each year, which adds to their value in the form of rental fees, as well as additional income of consumed wines and local products. Furthermore, other local suppliers, such as those operating natural and cultural heritage sites, benefit from retreat guests. Typically, the guests are highly aware of nature and are escaping from mass tourism, noise and polluted air.

They enjoy the unspoiled nature, the sound of bird songs in the morning, sunsets and starlit skies in the evening. Some providers have also prepared special culinary packages for cuisine lovers: while staying in the vineyard retreat each night the guests are taken to a different guesthouse, restaurant or farm tourism, where they can taste the diverse offer of Slovenia's cuisine.

To enhance market visibility, the Winegrowers Association hosts different bloggers and reporters, while also being active on social media (Facebook, Instagram), although they believe that a satisfied guest is by far the best promotion for them. They are very proud of the ratings higher than 9 on world-known platforms, such as Booking. com, which reflects the outstanding satisfaction of their visitors and guests. Guests compliment the beautiful nature, warm hospitality and cleanliness.

Usually the host will invite the guests to dinner, where they are briefed about local offers, cuisine, and are recommended interesting traditional events and tourist locations in the vicinity. Each guest is offered postcards with vineyard retreat images, which the hosts, on behalf of guests, will then take to the post office. This provides the opportunity for some additional low-budget promotion. The project started with Dolenjska and Bela krajina sub-region, then continued in Obsotelje and Kozjansko, and Lendavske gorice was added in 2019; it now covers one-third of Slovenia.

Conclusion

It is necessary to point out that the "Tourism in Vineyard Cottages" as innovative tourist package was a major step forward and proved that systematic work and cooperation could assist Dolenjska region in becoming more recognisable in this segment of its tourist industry.

Nevertheless, additional research into tourist markets, more coordinated activities and approaches are required for the successful future development of wine tourism in Dolenjska. It is important to integrate and unite all stakeholders in the tourist industry, private, public as well as the enthusiasts, in order for them to work together in a coordinated manner. This includes cooperation of wineries, hotels, tour operators, restaurants, wine cellars, farmers, food producers and wine merchants. Such an approach should be supported by research, analysis and assessments on how wine tourism can be integrated into sustainable tourism products. The main objective should be how to turn this into a convincing and marketable tourist product (Koščak and O'Rourke, 2020).

It is likely that the tourism of the past decades, the mass movement of tourist invasions, will be an image of the past. Also from the perspective of the current Corona pandemic, mass tourism was nothing else but an open wound in many national economies on the European continent permanently reinfected by hordes of consumer tourists. Mass tourism demands drastic changes and fundamental rethinking. The future will show whether new forms and proposed long overdue change to a scientific, human, rational and conservative niche tourism will be the single, practical cure for our collapsing tourism industry. And the product "Vineyards retreats" matches exactly all elements of the niche tourism; for Winegrowers Associations it is the future of sustainable tourism in Europe and world-wide.

From the last evidence, it is clear that this product and offer also has suffered the effects of the COVID-19 pandemic; statistics for 2020 showed that there were around 30% fewer visitors compared with previous seasons. It was also recorded that foreign guests were in the minority and that most likely the domestic market was and will be the most important in the 2020–2021 tourist season (Kompas Novo mesto, 2020).

References

Carlsen, J. and Charters, S. (eds) (2006) *Global Wine Tourism: Research, Management and Marketing.* Wallingford, UK and Cambridge, MA: Cabi.

Getz, D. (2000) *Explore Wine Tourism: Management, Development and Destinations.* Cognizant Communication Corp.

Hall, C. and Mitchell, R. (2002) The touristic terroir of New Zealand wine: the importance of region in the wine tourism experience. In Montanari, A. (ed.), *Food and Environment: Geographies of Taste*, pp. 69–91. Rome: Societa' Geografica Italiana.

Kivela, J. and Crotts, J.C. (2006) Tourism and gastronomy: gastronomy's influence on how tourists experience a destination. *Journal of Hospitality & Tourism Research*, 30(3): 354–377.

Kompas Novo mesto (2020) Internal statistical data of Vineyard Retreats marketing association.

Koščak, M. and O'Rourke, T. (2020) *Ethical and Responsible Tourism: Managing Sustainability in Local Tourism Destinations.* Abingdon, UK & New York, USA: Routledge.

Novelli, M. (2005) *Niche Tourism: Contemporary Issues, Trends and Cases.* Oxford: Butterworth-Heinemann.

Slovenian Tourism Strategy 2017–2021, Government of the Republic of Slovenia. Retrieved on 25 May 2020 from: https://www.slovenia.info/uploads/publikacije/the_2017-2021_strategy_for_the_sustainable_growth_of_slovenian_tourism_eng_web.pdf.

VisitDolenjska (2012) Retrieved on 25 May 2002 from: https://www.visitdolenjska.eu/en/.

Yuan, J., Jang, S., Cai, L.A., Morrison, A.M. and Linton, S.J. (2005) An analysis of wine festival attendees motivations: a synergy of wine, travel and special events? *Journal of Vacation Marketing*, 11(1): 37–54.

Case Study 3

Venice: the silent city

Introduction

This case study examines the situation in Venice, a UNESCO city. Initially, we will consider the situation in Italy as a whole before moving to discuss the specific situation in Venice. This is important in considering the overall economic situation as it affects tourism in Italy, given the importance of the country as a major tourism destination. It is useful to first consider the wider picture in Italy before turning to specific issues affecting Venice and the Veneto region. The issues to be looked at are tourism flows and values as well as the economic impact the COVID-19 virus has had and will continue to have on the Italian economy in general and the tourism sector in particular. [*LINK: Chapter 1/pp. 11–12*]

The previous situation in tourism

Data published by the UNWTO (UNWTO, 2018) indicated that Italy was the fifth top global destination by international arrivals and the sixth top destination by earnings from international tourists (UNWTO, 2019). The data showed that the total number of international tourists arriving in Italy in 2018 stood at 62.2m, which represented a 42.4% increase over 2010, thus an annual average increase of 5.3% (authors' estimates). Table CS3.1 indicates the breakdown of international tourism expenditure in Italy by type of tourist (overnight or day only).

Table CS3.2: shows the EU27 data for international tourism expenditure.

Table CS3.1 Tourism expenditure in Italy (2018)

Categories	EURm	Percentage of total
Overnight visitors	44,815	93.1%
Day visitors	3,334	6.9%
Total	48,148	100.0%

Source: https://ec.europa.eu/eurostat/web/products-statistical-reports/-/KS-FT-19-007

Table CS3.2 Tourism expenditure in EU27 (2018)

Categories	EURm	Percentage of total
Overnight visitors	334,006	89.5%
Day visitors	39,343	10.5%
Total	373,349	100.0%

Source: https://ec.europa.eu/eurostat/web/products-statistical-reports/-/KS-FT-19-007

Day visitors to Italy provide a relatively small percentage of expenditure compared to those who visit Italy for overnight stays and the figure given for their expenditure is lower than the EU27 average. Given this is data for the whole of Italy, it fails to represent the day visit phenomenon which relates to Florence, Naples, Rome and Venice. Finally to understand Italy's role in the EU tourism scene, Table CS3.3 indicates Italy's share in the percentage of EU27 inbound tourism expenditure in 2018.

It should be noted that in 2018, Italy's population was equivalent to 13.5% of the EU27 population and 13.1% of the EU GDP. It is also important to understand the economic value of tourism (including travel and hotels) within the overall Italian economy. According to projections (authors' data) Italy's positive balance of tourism in 2018 represented 19% of the EU27 positive balance. This suggests that in economic terms, in 2018 Italy was generating tourism activity well above its population and economic growth share of the EU27. Clearly tourism is also an important component within the Italian economy, representing 3.9% of domestic supply and supplying EUR 87.8bn of gross value into the Italian economy. In 2018, tourism created 4.2 million jobs representing 3.2 million Full Time Equivalent jobs.

The losses generated by COVID-19

Italy was impacted very early by the COVID-19 virus; despite having shut down flights from China, the virus appeared to have spread rapidly in Lombardy, the region hosting Italy's largest

Table CS3.3 Italy – share of EU27 tourism expenditure (2018)

Categories	Percentage of total
Overnight visitors	13.4%
Day visitors	8.5%
All international tourism expenditure	12.9%

Source:https://ec.europa.eu/eurostat/web/products-statistical-reports/-/KS-FT-19-007

airports (Malpensa, Linate, Bergamo), as well as being the heartland of Italian manufacturing. The lockdown was drastic and included the immediate closing of most industrial production as well as retail capacity. However, as it implemented the lockdown, the Italian government also introduced special measures for tourism and travel, including:

- Special financial allowances for those working in tourism and hotels to mitigate income loss
- Extending the social protection and social safety net to seasonal employees in tourism and hotels
- Financial support for tourism and hotel businesses
- Provision of guarantees for a refund voucher scheme for travel, tourism and hotels
- Provision of EUR 200m for the technically bankrupt airlines Alitalia and Air Italy

Despite these measures for enterprises and individuals in the tourism sector, catastrophic losses occurred in the tourism industry during the lockdown period and into the start of the recovery period from early June onwards. The losses to Italian tourism as a result of COVID-19, in both tourism inward numbers and income may be illustrated by the latest official data (authors' projections on unpublished data). Table CS3.4 summarises these losses up to 24 October 2020.

Italy operates all-year round tourism, as a result the lockdown affected the end of the Alpine season (skiing and related tourism activities) as well as the regular flow of early Spring tourists visiting

Table CS3.4 The impact of COVID-19 on tourism in Italy

Losses created March to October 2020	*Data*
Estimated number of visitors lost	216 million
Visitors lost as percentage of annual total	25.0%
Hotels nights lost as percentage of annual total	22.0%
Estimated expenditure of foreign tourists lost	EUR 25bn
Expenditure of foreign tourists lost as percentage of annual total	22.0%
Tourist income lost as percentage Italian trade balance	83.0%

Source: Authors' estimates based on Italian and EU data

Italian cities and the normal flow of visitors to Rome for the Easter religious period.

The future situation

The economic situation in Italy, despite the slow and halting return of some elements of tourism during June to August 2020, appeared to be poor. In the third quarter of 2020, GDP decreased by 4.3% on a year-on-year basis. This was propelled mainly by the collapse of both export trade and tourism, as well as the shut off of manufacturing capacity; at the same time annual inflation (HCIP method) as at October 2020 stood at -0.6% (Euro Zone figure was -0.3%).

The future prospects for Italy are indicated in Table CS3.5, which displays Eurostat economic analysis (Eurostat, 2020) for 2019–2021 in terms of the GDP quarter-on-quarter growth percentage change compared with the EuroZone.

An important element is the Current Account balance as a percentage of GDP, as tourism expenditure by foreign tourists is an essential component of this economic measure, as indicated in Table CS3.6.

Table CS3.5 GDP quarterly percentage change Eurozone/Italy

Years	Italy	EuroZone
2019/q3	0.5%	1.4%
2019/q4	0.1%	1.0%
2020/q1	-5.6%	-3.3%
2020/q2	-17.9%	-14.7%
2020/q3	-4.7%	-4.3%

Source: Authors' projections based on Eurostat and Italian national data, 2020

Table CS3.6 Current Account Balance as percentage of GDP

Years	Italy	EuroZone
2019	3.0%	2.7%
2020	3.1%	2.6%
2021	3.0%	2.7%

Source: https://data.worldbank.org/indicator/BN.CAB.XOKA.CD

This then affects Italy's Balance of Payments which feeds into GDP real growth. As we can see, the Current Account Balance remains relatively static as a percentage of GDP. But importantly, the GDP growth is predicted to crash fairly decisively in 2020 to -9.1%, which is significantly below the EuroZone GDP growth figure of -7.5%.

We should also note that although predictions of recovery show Italian GDP growth at 4.8% in 2021 (close to the EuroZone GDP growth figure of 4.7%), this will fail to make up for the drastic economic shock of 2020. Indeed, assuming a standard scenario of where economic activity picks up by October 2020, this will ensure that economic growth in real terms will not reach the 2019 level until possibly into 2022/2023.

Venice: before COVID-19

First we should consider the data which describes the situation in Venice prior to the lockdown due to the dissemination of COVID-19. It is estimated that tourism numbers to the historic city stood at 8.5 million by the end of 2019 (authors' estimate based on statistical data from the Comune di Venezia). Of these around 2.1 million were cruise passengers participating in city tourism for a day visit. Across Italy as a whole, day visitors comprised 6.9% of total tourism visitors; in the historic city centre (*centro storico*) of Venice, cruise passengers accounted for 25% of the inbound tourists alone. However, the number of day visitors in 2019 is estimated at 40% of all tourists in the *centro storico*, which would suggest that around 15% are non-cruise passengers, but are arriving as coach parties or as individuals arriving by train, bus or car. Whilst the day-visit phenomenon is common across Europe's historic cities, in Venice and Dubrovnik (see Case Study 1) it is concentrated into a relatively small space. This concentration may explain the density of day tourists as shown in Table CS3.7.

Table CS3.7 Density of day tourism

Location	Index*
Italy	1
Veneto region	3
Comune di Venezia	512

Source: https://www.comune.venezia.it/it/content/studi

Note: *number of tourists per km^2

The Veneto region is relatively large, but with a high population in the urban areas; nonetheless the density of tourism population is quite staggering and explains the concerns of environmental groups within the city about the extremely high volume of visitors to the city. It is therefore important to understand the key elements in regard to this situation in Venice. These are:

- The effects of over-tourism in the last decade
- The flooding of November 2019
- The impact of the COVID-19 virus on the city
- The economic impact of the COVID-19 virus

To understand the background of how Venice has come to this present situation and where this might lead in future, we need to analyse the following four components.

Over-tourism

The key issues are depopulation, degradation of the heritage and culture, and immense pressures on the costs of accommodation. These factors are driven by a lack of reasonably priced rented accommodation, excessive transits by cruise ships, and the destructive effects of day tourism. The Venetians have a dialect phrase to describe the day visitors, especially those from cruise ships – *"mordi e fuggi"* – which translated means "bite and run". This categorises the thousands of visitors who simply arrive, grab a snack or an ice-cream, and then run back to their cruise ship or tourist coach. Their commitment to the culture and heritage is fleeting and ephemeral – it is simply adding Venice to the check list of locations visited. As a result, over-tourism [*LINK: Chapter 1/pp. 14–15*] may be specified in the following ways.

- Mass numbers of day tourists spending little in real terms – for example at the height of the tourist season (April to October) the authors estimate that 32,000 cruise passengers per day normally visited the *centro storico* – representing 58.2% of its population. Unlike land-based day visitors (e.g. coach trips from neighbouring cities or tourists travelling within the local area), we would suggest that cruise passengers spend a relatively small amounts during the period they are in the city, as they receive accommodation, meals and refreshments on the cruise ship. Thus even the relatively low monetary income that they may generate is easily outstripped by the environmental damage that ships and their passengers cause to the cultural and physical heritage, as well as the cost of the waste they generate.

- Large-scale upward pressures on the price of accommodation for local inhabitants – the most potent example of this is the effect of Airbnb. Recent data (www.insideAirbnb.com/venice, 2020) estimates that 12% of the housing stock in the *centro storico* is utilised for Airbnb accommodation, which does not include privately owned and let apartments or apartments owned by tourism companies. It is estimated that letting an entire apartment will generate earnings of EUR 19,224 per annum. In the *centro storico*, 75% of Airbnb lettings are for entire apartments rather than for a room in an apartment (which was the original concept of the online service – to give visitors a taste of living with local people). In July 2015, 3,000 apartments were listed by Airbnb in the *centro storico*; in July 2019, this had leapt to 8,800 (i.e. 193%). This has taken out a huge slice of the local rented accommodation sector and in particular has forced young people as well as students at the Ca'Foscari University to live on the mainland in Mestre or surrounding towns. We should also take account of the fact that two-thirds of AirBnB hosts have multiple listings, in other words they market more than one apartment. However, at the same time, it is noteworthy that with a potential 12 week shutdown of Venice due to the corona virus, Airbnb renters will be losing approximately EUR 4,400 of income.

- Population shift – According to official data (Comune di Venezia, 2018), between 2003–2015, the population of the Comune (city, plus Mestre and islands in the lagoon) fell by 13%, whilst the number of visitors rose by 64%. Between 2016–2019, the authors estimate that visitor numbers rose by 25.2%; in that same period the population of the *centro storico* is estimated to have fallen by 14.6%. We estimate that the current population (February 2020) of the *centro storico* is approximately 53,000, of whom around 13,700 (26%) are over 60 years of age. In fact, the current population is equivalent to the estimated population in 1020 and is 26% of the population in 1600; very generally most global historic cities (e.g. Cairo, Montreal, Paris) have expanded dynamically over the last millenium. Importantly, only 7,400 (14%) are under 18 years. Furthermore, due to the high purchase costs of housing, local inhabitants are unable to compete, especially with purchasers from outside the EU. The ageing population would also indicate, that unless price pressures diminish in the wake of the corona virus and potential recessionary trends, there will be a constant flow of housing for sale as the elderly either die or decide to leave the city.

Flooding

From 12–17 November 2019, the city endured the most catastrophic flooding from the rise of waters in the lagoon (the *"acqua alta"*) since 1966. At one point the tide reached 1.87 metres, boosted by strong Alpine winds from the north. This resulted in a total of 80% of the *centro storico* being underwater in excess of one metre in most cases; for example 60 churches were damaged by the flood water, including the Basilica of San Marco, which underwent serious flooding. The cost of basic structural repairs to damaged property has been estimated at EUR 400m.

Whilst such damage to the historical and cultural heritage was immense, with the closure of museums and cultural institutions, there was also an immediate socio-economic effect on facilities such as hotels, restaurants, artisan workshops and shops. In addition, as many of the less affluent inhabitants (e.g. students and workers in tourism) live in ground floor accommodation, many of those properties were badly affected by such a level of flooding. At the time, it was estimated that the immediate repairs involved would have lasted at the minimum to around May 2020. The sheer cost of repairs, with the financial strains implied by the COVID-19 pandemic, has also placed a further disincentive on pro-active measures to halt the declining population of the *centro storico*.

The virus strikes

Due to cases of corona virus in the Veneto region (given its proximity to the epicentre in the Lombardy region), Venice began to suffer a reduction in tourism flows from late February 2020 onwards. Within weeks the regional and Italian governments had effected a lockdown of the city with the consequential total flight of tourists. Therefore Venice had moved from over-tourism to no tourism; if this is added to the physical and financial effects of the November floods, the flight of tourists and the restriction of the inhabitants to their houses resulted in a silent city already reeling from the November disaster. Obvious sectors were impacted – cafes, bars, tourist attractions – which in turn impacted on ancillary occupations such as tour guides, craftsmen and artisans whose whole livelihoods are connected to tourism. There were also a number of serious cultural issues – the February Carnival was cut short, the Biennale of Architecture was postponed and all museums and cultural events closed.

Table CS3.8 indicates the spread of the COVID-19 virus in the Province of Venice – this is the city of Venice and the immediate mainland areas around the lagoon. It is clear that similar to the overall Italian situation, the pandemic accelerated rapidly through until May 2020 and then declined over June and July 2020 before accelerating through August into September 2020. It is clear that the city has been strongly affected by the virus in terms of the shutdown of tourism inflows, but an appropriate question is whether this is now a time for Venice to pause, reflect and then reset, regarding its tourism inflows and balancing these against the environmental damage created.

The local economic impact of COVID-19

The authors estimate that the Veneto region produces 9% of Italian GDP; this includes tourism not only in Venice itself but in the surrounding region with such historic cities as Padua, Rovigo and Vicenza – the historic *terra firma*. To an extent these surrounding historic locations feed into and feed from tourism in the city; in cost terms, tourism stays in Padua, Rovigo and Vicenza are significantly less expensive in terms of accommodation and living costs than in the centre of Venice. At the same time, tourists in Venice will gravitate towards those three cities due to their artistic, architectural and cultural connections with Venice. The rural areas in the *terra firma* are also centres for responsible and

Table CS3.8 Progress of COVID-19 in Provincia di Venezia (total infections on date)

Date	Provincial monthly growth rate	Italian monthly growth rate
24.03.20	-	-
24.04.20	179.6%	179.0%
24.05.20	13.2%	19.1%
24.06.20	1.2%	4.2%
24.07.20	3.5%	2.6%
24.08.20	11.8%	6.0%
24.09.20	22.4%	16.9%
24.10.20	71.1%	65.8%

Source: Base data http://opendatadpc.maps.arcgis.com/apps/opsdashboard/index.html#/ b0c68bce2cce478eaac82fe38d4138b1 augmented by authors' calculations

sustainable agro-tourism, but at the same time visitors to such facilities will also have an interest in visiting Venice itself. The lockdown therefore has had an impact on these ancillary districts outside the historic city itself in terms of flows into the city from the urban and rural locations, but also in regard to trips from the city out to these districts. Furthermore, we also have to take account of the goods and services production to support the main tourism centres. The *terra firma*, particularly in the Rovigo area, has been a major supplier of fruit, vegetables, cheeses and local meats to Venice from the 11th century onwards. According to data from the city authority (Comune di Venezia, 2019) supplemented by the authors' projections, we estimate that the total number of individuals staying in hotels and similar establishments in the *centro storico* alone in 2019 would have reached 5 million. To provide fresh food items from the surrounding countryside constitutes a major business activity, and the cessation of tourism has resulted in a major loss of income for those producers.

Conclusion

For decades, environmentalists have campaigned against the depredations of the Marghera oil refinery, the cruise ships and the masses of day visitors. Suddenly in March 2020, that environmental problem was partly resolved; there were no massive cruise ships polluting the Giudecca and no tourists polluting the city streets. It was Day Zero, the silent city, bringing memories of 1,600 years ago when the refugees of the Celtic Veneti tribe, fleeing the mainland from the Huns and the Lombards, settled on the muddy islands of the Venetian lagoon.

But this now raises a complex problem; if Venice rebuilds tourism, will it seek to return to the previous over-tourism model. Indeed this model was already being contested by the city authorities, supported by environmental campaigners, in terms of restricting cruise ship passages through and into the lagoon and reducing the number of day visitors to the *centro storico*. Will the city be able to restrict tourism activity, as it continues to re-open, to responsible, sustainable and ethical tourism actions? Mass tourism provided income at a macro-level in terms of fees through the city port from the cruise ships and the tourist coaches arriving daily at the Piazzale Roma. But set against this was the environmental damage to the lagoon, to the buildings of the historic city and the pricing pressures which pushed out city inhabitants. This is clearly an interesting conundrum to solve. But there is hope; over the past two decades the voices of those seeking a more honest and responsible form of tourism for the city have increased and have begun to gain some degree of success despite Italy's byzantine political system.

An emboldening example is the social enterprise "Venezia Autentica" (www.veneziaautentica.com, 2020), whose mission it is to halt the exodus of residents by proposing an alternative to mass tourism, making it easier for travellers to enjoy a more meaningful and authentic experience of the city whilst making a positive impact on the local community, environment, and economy. The enterprise's goal is to transform the way tourism impacts the city by moving from unsustainable mass tourism to a more sustainable and responsible tourism. Until the corona virus hit, they were using social innovation and digital technologies to enable visitors to have a quality experience in Venice while making a positive impact locally.

One might hope that the new tourism situation which emerges in the future will give less weight to tourism numbers, but greater importance to responsible and ethical tourism. This, essentially, has the potential to bring higher monetary and social rewards to the local community than the simple processing of millions of mass tourists who visit the city for some form of cultural and historical voyeurism. In a sense, the post-virus restoration of tourism, provides a massive opportunity for us to shape ethical and responsible tourism in Europe's historic cities that for decades have been polluted by unrestrained and irresponsible tourism.

References

Citta' di Venezia (2018) *Annuario del Turismo 2017*, Assessorato al Turismo, Venice.

Citta' di Venezia (2019) *Annuario del Turismo 2018*, Assessorato al Turismo, Venice.

UNWTO (2020a) *All Countries: Inbound Tourism: Arrivals 1995–2019*, Tourism Statistics, UNWTO, Madrid.

UNWTO (2020b) *All Countries: Inbound Tourism: Arrivals 1995–2019*, Tourism Statistics, UNWTO, Madrid.

Case Study 4

Malta – regaining a sustainable heritage

Introduction and background

The island of Malta, an archipelago of three islands, is located some 120 km south of Sicily with the coast of Tunisia about 300 km distant to the south-west. As a case study it is of interest due to the fact that the island's tourism profile contains a mix of mass tourism. This covers resorts with large hotels and high-occupancy apartment blocks, upmarket resorts with luxury hotels and marinas, self-catering villas outside towns, and other forms of self-catering accommodation. In addition there is an element of cruise tourism, although the numbers are relatively limited in comparison to other mediterranean destinations such as Barcelona, Dubrovnik and Venice. This case study therefore examines the pre-March 2020 condition of tourism in Malta, and then evaluates the result of the COVID-19 pandemic as well as the consequent economic downturn which evolved from April 2020 onwards. Tourism represents 16.7% of Malta's GDP based on data provided by the Malta Tourism Authority (MTA) and the National Statistical Office of Malta (NSO) (MTA, 2020; NSO, 2020). The sector employed 22,866 individuals in 2019, of whom 58.4% were full-time and 41.6% part-time (MTA, 2020). The full-time tourism employees constituted 5.2% of the workforce; it is estimated 38% work in the hotel and accommodation sector and 62% in other tourism activities including restaurants, cafes and tourism services. Total inbound tourism (which excludes overnight cruise passengers) is shown in Table CS4.1.

The slower level of growth in 2019 over 2018, which also resulted in a 6.4% drop in visitors from Scandinavia, Germany and Austria may in part be explained by the collapse of the German-owned Thomas Cook Group for which Malta was an important destination. Tourist expenditure is indicated in Table CS4.2.

Table CS4.1 Inbound tourists 2017–2019

Year	Numbers	Year-on-year change
2017	2,273,837	
2018	2,598,690	14.2%
2019	2,753,329	5.9%

Source: https://www.mta.com.mt/en/file.aspx?f=32328

Table CS4.2 Tourist expenditure 2017–2019

Year	EUR Millions	Year-on-year change
2017	1,946.9	
2018	2,101.8	7.9%
2019	2,220.6	5.7%

Source: https://www.mta.com.mt/en/file.aspx?f=32328

Table CS4.3 Tourist expenditure 2017–2019 per capita

YEAR	EUR PER CAPITA
2017	856
2018	809
2019	807

Source: https://www.mta.com.mt/en/file.aspx?f=32328

A further calculation is the per capita expenditure of inbound tourists, given in Table CS4.3.

Structure

The average length of stay over the period 2017–2019 was 7.1 nights (authors' estimates based on MTA data). The Maltese tourism product does display a degree of seasonality but it is not as summer oriented as other South Mediterranean destinations. According to MTA data for 2019, 34% of tourists visited in July–September, 28% in April–June, 23% in October–December and 15% in January–March (MTA, 2020). Similarly tourist accommodation occupancy rates were only above the annual average of 66% in April–June (74%) and July–September (83%). The largest increase by quarter in 2019 (year-on-year) was in the October–December period which grew by 10.0%, over the 5.9% total increase in tourism. In part this may reflect increasing numbers of Northern European tourists availing of discounted rates for 1–2 month stays over the October–December period.

Examples of the range of accommodation display a range between 5* de luxe hotels (large international style as well as luxury resorts and boutique hotels) which represent about 13% of accommodation compared to guest houses representing about 8%. Self-catering accommodation, which includes purpose-built multi-storey apartments as well as country villas represents around 29% (MTA, 2020).

The origins of tourist vistors to Malta in 2019 are indicated in Table CS4.4.

Given an average spend of EUR 807 per inbound tourist, only German visitors appear to have exceeded the average on the 2019 data. Although the UK is the biggest market, as the former colonial power, the proximity to Italy (and a shared cultural history) ensures Italians are the second highest group of tourists. English is the second language, whilst Italian is understood by around two-thirds of the Maltese population.

Table CS4.4 Origin of tourists 2019 – top 5 originating countries

Year	Number	Percentage of total	Average spend
UK	649,624	23.6%	EUR 792
Italy	392,955	14.3%	EUR 602
France	239,140	8.7%	EUR 804
Germany	211,546	7.7%	EUR 830
Spain	116,295	4.2%	EUR 654

Source: https://www.mta.com.mt/en/file.aspx?f=32328

Another important factor is the distinction between tourists who arrive on a holiday package (flight/accommodation/associated costs) and those who book travel and accommodation separately. In 2017, tourist expenditure on package holidays amounted to 44.9%, whilst those purchasing flights and accommodation separately was 55.1%. In 2019, the package holiday element was 39.7%, with non-package at 60.3%. At the same time, in 2019, package tourists contributed expenditure of EUR 683 per capita as compared with non-package tourists at EUR 483 per capita.

Cruise passenger traffic is often regarded as a major driver in overtourism. In Malta, given there is no significant day-trip tourism (apart from very minimal flows from Sicily on the passenger only ferry service), cruise passengers constitute the only important day tourism flows. In 2019, the total number of passengers landing in Malta was 765,696 – of whom 82.2% were transit passengers – i.e. they were leaving or joining cruises via the Malta International Airport. The number actually overnighting in Malta in transit was 18,649. Whilst these passengers tended to take day trips to tourism-focused features in Valletta and Mdina, their impact was primarily during the 10:00 to 16:00 period. The number of cruise ships calling at Malta in 2019 was 359 (up 5.0% on 2017). The estimated density of cruise ship passengers across the entire territory of Malta is calculated at 431 per km^2. It is critically important that in 2020 total cruise passengers into Malta stood at 47,193, a decrease of 92.1% year-on-year. Total cruise passenger traffic during the third quarter of 2020 amounted to 7,018 – a year-on-year decrease of 97.3%.

We should also consider the balance of tourism in expenditure terms; in 2019 with EUR2,221m earned from inbound tourism, the outbound flows from Malta accounted for EUR606m – a positive balance of EUR1,614m. The structure of tourism on the two islands which comprise Malta (Malta mainland and Gozo) indicates that in 2019 there were a total of 55,597 beds of which 62% were in hotels or related accommodation and 38% in self-catering accommodation.

Malta and COVID-19

Due to the nature of an island environment, Malta experienced a relatively low number of COVID-19 cases from the beginning of the pandemic. During April to May 2020, Malta ranked among the bottom five of European Union figures for infections and death rates. In part this was a result of the closing of the island from air and sea transport entry on 11 March 2020, which immediately stopped the flow of tourism. Malta enacted further controls over tourists entering from any country on 13 March 2020, requiring that they must spend 14 days in mandatory quarantine. On 20 March all passenger flights inbound to Malta were suspended and the airport was closed other than for humanitarian or emergency repatriation flights; this travel ban lasted until 31 May 2020. The opening of the island to tourism on 1 July, although not having an immediate effect on the number of infections, did however result in a secondary spike that became apparent into September 2020. By the end of September 2020, Malta ranked seventh in the EU/EEA countries in terms of the number of COVID-19 cases per 100,000 inhabitants. In fact, between 31 March to 31 October the infection rate grew by 3416% and at the end of October Malta ranked 16th in the EU/EEA countries.

The effect on tourism

The effect of the ban on travel, closure of accommodation, and the inability of potential tourists from the key markets (UK, Italy, France, Germany and Spain) to travel out of or back into their countries had a profound influence on Maltese tourism in the second quarter of 2020. January and February 2020 displayed a growth in tourist arrivals over the previous year. In January 2020, tourist arrivals were up 18.8% over January 2019 and in February 2020 up 13.6% over February 2019 (Eurostat, 2020). We estimate, based on national data, that the closure of the island to tourism had the effect of reducing annual tourist numbers by 28.6% giving an estimated financial loss of EUR635.1m. Based on the GDP figures for 2019, this would have represented a loss of tourist income, calculated as a percentage of GDP, at 4.8% (MTA, 2020; NSO, 2020).

However, urgent steps were taken to restart the tourist economy. During May 2020, the Maltese government in collaboration with the tourism industry established a set of procedures for a number of tourism establishments and other tourism receiving operations to ensure rigid compliance in the fields of social distancing, enhanced hygienic practices and the minimisation of COVID-19 infection risks. This was to ensure the potential for re-opening of tourism activities given the disastrous loss of national income due to the existing closure of facilities.

It is clear that Maltese tourism suffered significantly during the period of the initial lockdown, and subsequently. The National Statistical Office (NSO, 2020) reported in September 2020, that in the period January–July 2020, the year-on-year (y-o-y) figures for inbound tourists had fallen by 72.3%, tourism expenditure fell y-o-y by 77.9% and tourism expenditure per capita fell by 20.2% y-o-y. Taking July 2020 over July 2019 (July being one of the main tourism input months), the number of tourists fell by 84% and expenditure fell by 88%. Table CS4.5 indicates the January–June 2020 tourism per capita spending data in respect of the top spending countries, whilst Table CS4.6 indicates the same data in respect of the top five inbound countries by the number of tourists.

Table CS4.5 Tourism expenditure per capita – top spending countries (Jan. to June 2020)

Year	Average spend	Percentage of Malta inbound
Australia	EUR1118	0.6%
USA	EUR1014	1.8%
Switzerland	EUR815	1.5%
Austria	EUR767	1.1%
Ireland	EUR676	2.9%

Source: https://nso.gov.mt/mt/Pages/default.aspx

Table CS4.6 Tourism expenditure per capita – top countries by number of tourists (Jan. to June 2020)

Year	Average spend	Percentage of Malta inbound
UK	EUR590	22.0%
Italy	EUR493	13.5%
France	EUR619	8.2%
Germany	EUR675	8.0%
Poland	EUR496	5.8%

Source: https://nso.gov.mt/mt/Pages/default.aspx

The above data indicates tourists from the top spending countries – as opposed to those countries which send the largest number of tourists to Malta. This data appears to have been affected by the restrictions on arrivals from specific countries due to the COVID-19 pandemic.

In comparison, this table lists the top five countries in terms of the numbers of tourists whilst listing their spend per capita.

The effect on the wider economy

In 2019, Malta's economy showed a growth in real GDP of 4.4% (International Monetary Fund, 2020) against a EuroZone growth level of 1.2%. Importantly, Exports of Goods & Services (which includes tourism) stood at EUR15,579m against Imports of Goods & Services of EUR13,492m – a trade balance of EUR2,087m. The Current Account balance as a percentage of GDP was as indicated in Table CS4.7, displaying that it is well above the EuroZone average:

Table CS4.7 Malta's current account balance as percentage of GDP (with comparison to the EuroZone)

Year	Malta	EuroZone
2019	8.4	2.7
2020*e*	3.3	2.6
2021*p*	6.1	2.7

Source: https://www.imf.org/en/Publications/WEO/Issues/2020/04/14/weo-april-2020

Note: e=estimated; p=projected

Further more recent data (NSO, 2020) in Table CS4.8, provides information about the actual Q1 Exports of Goods & Services between the first quarter of 2018–2020, of which tourism is a component:

Table CS4.8 Malta's Exports of Goods & Services Q1 2018–2020

Quarter 1	EURm	change y-o-y
2018	3740.2	4.2%
2019	3844.6	2.8%
2020	3662.2	-4.7%

Source: https://nso.gov.mt/mt/Pages/default.aspx

Table CS4.9 Malta's real GDP growth 2019–2021 (with comparison to the EuroZone)

Year	Malta	EuroZone
2019	4.4	1.2
2020e	-2.8	-7.5
2021p	7.0	4.7

Source: https://nso.gov.mt/mt/Pages/default.aspx

Note: e=estimated; p=projected

Table CS4.10 Malta's real GDP growth Q1 2018–2020 year-on-year (with comparison to the EuroZone)

Year	Malta	EuroZone
2018	5.5%	0.3%
2019	6.2%	0.5%
2020	0.5%	3.8%

Source: https://nso.gov.mt/mt/Pages/default.aspx

However, the effect of the dramatic loss of tourism income as well as other macro-economic effects concerning this island economy indicates a significant decline in national economic growth, as displayed in Tables CS4.9 and CS4.10.

Conclusion: the balance between mass tourism and sustainable tourism

As with many tourism destinations which are also islands, an island location brings problems of its own. It is not possible for the majority of visitors to enter by environmentally friendly methods (e.g. by train, on foot, or by using sea vessels which are environmentally friendly). Malta with a size of 316 km^2 and a population of 475,000, has a density of 1,503 per 1 km^2; the majority are resident in the capital Valletta and its suburbs. Adding 2.7 million tourists (excluding cruise ship visitors), the majority of whom stay for seven nights, imposes a heavy environmental strain on the national infrastructure, as well as emissions from road traffic, shipping and over 4,000 aircraft movements in an average month alone.

It may be suggested that Malta is now at a unique tipping point; it has been evident for some time that mass tourism has affected some elements of Maltese life and culture. In the northern resorts of Buġibba and Qawra, large tourist hotels, pools and beaches aimed at families jostle side-by-side with the "stag & hen" parties of drunken northern European tourists. The east coast marina town of St Julians Bay, home to many of the international chain hotels (Hilton, Intercontinental, Radisson etc.) has in the past few years become a building site for the construction of new tower blocks. In the ancient city and former capital of Mdina, from mid-morning the bus terminal and parking space outside the city walls is thronged with tourist buses and the ubiquitous cruise ship passengers seeing Malta in a day. At the same time, tourists in hotels each produce 1.25 kg of waste daily, compared to one person in a Maltese household producing 0.68kg.

Clearly there are significant pressures at a global and EU level to develop higher levels of ethical and responsible tourism, managed in a sustainable way at a local level, despite the economic benefits that tourism brings. Dr Gabriel Farrugia, Projects Officer of the Malta Tourism Society has suggested that in the case of Malta an integrated and sustainable approach engaging a wide number of stakeholders should be developed as part of the post-COVID 19 strategy. The stakeholders would be tourism professionals, academics and communities; this would then lead into community-based tourism experiences engaging visitors with the rich variety of local life in all aspects. Dr Farrugia also suggested that

> "we must at the same time strive to protect our natural and built heritage not for the sake of the mass visitor, seen as a necessary evil one must live with to keep up the economic growth, but for the sake of preservation and posterity as well as for cultural education".
>
> (Malta Tourism Society, 2020)

References

Eurostat (2020) Tourism arrivals 2019. Retrieved 31 January 2021 from: https:// ec.europa.eu/eurostat/statistics-explained/index.php?title=Tourism_statistics

International Monetary Fund (2020) *World Economic Outlook Update, June 2020*. Washington DC: IMF.

Malta Tourism Authority (2020) Malta, facts & figures, 2019, Valetta: MTA. Retrieved 31.01.21 from: https://www.mta.com.mt/en/file.aspx?f=34248.

Malta Tourism Society (2020) Verbal evidence.

National Statistics Office – Malta (2020) https://nso.gov.mt/mt/Pages/default.aspx.

Case Study 5

Logarska valley in Solčavsko region, Slovenia

The Solčavsko region is located at the upper current of the Savinja River along the Slovenian–Austrian border. The area is surrounded by a mountain chain consisting of the Kamnik-Savinja Alps and Karavanken Alps. Solčavsko consists of three Alpine glacial valleys: Logarska dolina, Robanov kot and Matkov kot. In the middle of the area lies the small village of Solčava with several dozen inhabitants, whilst the whole municipality numbers only about 600 inhabitants. The area comprises 103 km², but there are on average only five inhabitants per km². More than 80% of the area is environmentally protected with two landscape parks (Robanov kot and Logarska dolina) and Natura 2000 Network – a network of nature projection areas within the EU. In such an isolated settlement there is sufficient space for the more than 100,000 annual visitors (with about 20,000 overnight stays), who visited the region in the period from 2015–2019. However, the carrying capacity of some of the more frequently visited places and the functioning of visitor management in a wider area must be taken into account in protecting nature and the cultural landscape. Tourism remains closely connected with traditional activities – agriculture, forestry and handicrafts – for centuries, the largest farms in this Alpine space have remained more or less self-sufficient.

Solčavsko is also characterised by its centre, Solčava, and by ecotourism. There are intentionally no large ski-lifts, only traditional activities. The resolute and persistent mentality of local inhabitants is a particular feature of Solčavsko, and of a community that is also cohesive and open. The proof is in the number of visitors who are drawn to the natural beauty and recreation possibilities that the region offers and also to the excellence of its facilities (Slapnik and Bogataj, 2016). The milestones of sustainable development began at the end of the 19th century with the breeding of indigenous Jezersko-Solčavska sheep and in 1950 led to the local initiative of "Robanov kot natural value protection". Decades later, in 1987, the landscape parks Logarska dolina and Robanov kot were established, and in 1992 the concession for nature park management in Logarska dolina was acquired by local inhabitants – an unprecedented event in Slovenia.

Subsequently, from 2005 onwards, international environmental institutions recognised the efforts of Solčavsko with awards from national and international organisations for park management, destination excellence, architecture and ecology. Such achievements are not easy due to the extremely heterogeneous nature of the Solčavsko region. Many inhabitants live on farms high in the mountains, others cope with the conditions in Alpine valleys such as Robanov kot and Logarska dolina. Furthermore, the central settlement of Solčava is different (buildings in a form of grouped village) from smaller settlements in the valleys and on isolated and scattered farms. Thus, the municipality and region copes with a diversity of development models; nonetheless in Solčavsko's past, present and future, the constant motive for all its development activities has been sustainability (Slapnik and Bogataj, 2016).

The area's development focus

Solčavsko, the areas of Logarska Valley and Robanov and Matkov kot have been the subjects of research on three occasions by international expert groups and students who jointly in 2007 prepared some key elements to determine the future of the area (Anko et al., 2007, pp. 104–106).

1. Natural resources and the environment

- studying and respecting natural conditions and historic adaptations to them
- developing management strategies to protect (and where appropriate, to market) natural resources such as water, wildlife, fish, minerals and mountain wilderness
- building awareness (amongst residents and visitors alike) of the value of the superlative natural environment

2. People

- fostering the education, skills, engagement and commitment of the future generation
- recognising that human and social capital are as important as numbers of people as indicators of the future
- developing appropriate municipal policies to implement the concept of sustainability

3. Farming

- recognising that future farming will be based not only on food and raw material production, but also on cultural landscape management – "living off the land" through farm diversification and multifunctional land management

- developing new education and training opportunities
- helping farmers to sell their products through local branding and better marketing.

4. Forestry

- adding as much value as possible to timber through local processing before export
- marketing the many forest functions (beyond timber production) under forestry law
- building awareness of the importance of these functions (including the role of mountain forests in nature conservation) among forest owners and general public

5. Tourism

- developing tourism as an important additional economic activity, for example to maximize the added value of domestic (farm) products
- understanding the negative potential of tourism and developing appropriate mitigation strategies, including building tourism as a process of cultural exchange and carefully planned "marketing" of natural heritage
- attracting an appropriate diversity of visitor types, with due regard to the needs of the area and of potential visitor groups

6. Potential Kamniško-Savinjske Alps regional park

- Developing a participatory planning process involving all "stakeholders', resulting in:
- an understanding of its strengths, weaknesses, opportunities and threats of Park status
- new employment and training opportunities for local people
- stronger protection for local biodiversity and natural features
- maintenance of the distinctive lifestyle and culture of the area

The area followed this strategic capacity under the management of the Local Development *Center Rinka*, based in Solčava of which a new visitor centre was part. Visitor management is a tool for sustainable tourism development in the natural environment; on this principle, the *Center Rinka* was established in the village of Solčava.

The multipurpose centre was built with the help of a Norway Grant in 2011 and it promotes local products (such as wood, wool and food), local people and their skills and knowledge. The building incorporates contents and functions important for the whole region – a tourist information point, a multimedia hall, permanent and temporary exhibitions, a business incubator and municipal offices. There are two main highlights within *Center Rinka:*

1   a permanent exhibition titled "Solčavsko—a Walk into the Lap of the Alps", emphasising local attractions and stories
2   an exhibition and selling space for local products. From the smallest photo to the largest patchwork, each artefact and every product in *Center Rinka* has its own story and is related to other items, stories and people

The area's popularity with domestic and international visitors drives the need to vigorously protect the nature and the local inhabitants of the area from over-tourism.

Logarska Valley

Not only growth in numbers but also changes in visitor types have caused tremendous strain, triggering one of the most important recent developments – the establishment of "Logarska dolina Ltd." (as a non-profit company) to protect the environment and interests of local landowners in Logarska dolina. The project was launched by a local inhabitant, Avgust Lenar, who provided a new governance model with the slogan "Logarska dolina is not here to be changed, but to change you" – symbolising a strong confidence in Solčavsko's own strengths. Since its establishment the company has managed different projects of sustainable development in Solčavsko; the shared vision of all the projects and the mentality of the local inhabitants is epitomised by the above slogan.

Lenar, a forester by profession but since 1992 the manager of Solčavsko's Landscape Park, describes the region's developmental path in the following terms:

"In the 1980s, farmers in Logarska dolina had several problems with many careless visitors parking in a natural environment, setting fires, littering and causing damage to the environment and locals in different ways. In 1987, the municipality created a Landscape Park. However, there were no control services or money for development. Therefore, the landowners established a company in 1992 and obtained a concession for the management of the park. They started to collect entrance fees for motor vehicles. It was the first such example in Slovenia where locals managed the park, which led to disapproval from visitors and the professional community. But the awards and wide recognition for the successful management of the protected area soon confirmed that the decisions of the locals had been correct." (Lenar, 2013)

The efforts of the local stakeholders and the unique management style of Landscape Park Logarska dolina have been supported by many experts. Among them, Boštjan Anko, professor at the Biotechnical Faculty, University of Ljubljana and his international team of experts, have greatly contributed to the development of the park and to the broader understanding of protected areas. In 2007, Anko wrote:

> For the last two or three generations, the locals of Solčavsko have seen their chance to survive and their future in self-sufficient farming, market-oriented livestock production, income from the forest, mass tourism, ecotourism … The whole future development should be seen as a collection of the best of what each of the above activities has to offer. Finding the future will not be oriented towards finding a single "ideal" activity, but towards finding a harmonious coexistence of all of these activities (which are part of Solčavsko identity) – and some more – on the basis of sustainable thinking. The stability and strength of Highland people and their farms lies in their conservatism and their mistrust towards novelties, which leave little room for manoeuvre for correcting mistakes in unfavourable natural conditions.
>
> (Anko et al., 2007, pp. 64–65)

Tourism and COVID 19 in Logarska Valley

As stated previously, the number of visitors has been growing constantly during the period from 2015–2019. In Table CS5.1 we present the most recent data from the period 2017–2020, which indicates visits in terms of transport mode.

These results display a constant flow of visitors, calculated on the basis of three passengers to a car, five in a minivan, 50 in a bus and two on a motor-bike. This suggests in the region of 11,000 visitors per year on the basis that these visitors are only those who paid the entrance fee (there are some others, i.e. those who have annual permits and those, who entered the valley out of the time when entrance control is operational) of

Table CS5.1 Number of visitors by transport mode in Logarska Valley, 2017–2020

Year 2020

Month	car	bus	mini van	motorbike
May	1,327	0	47	165
June	2,356	2	78	230
July	5,795	3	226	386
August	7,124	7	222	368
September	3,687	17	113	291
TOTAL	**20,289**	**29**	**686**	**1,440**

Year 2019

Month	car	bus	mini van	motorbike
April	124	4	4	0
May	1,767	25	45	89
June	2,959	64	125	431
July	4,506	15	247	325
August	7,727	21	376	527
September	2,324	43	121	239
October	1,256	18	57	68
TOTAL	**20,663**	**190**	**975**	**1,679**

Year 2018

Month	car	bus	Mini van	motorbike
April	1,307	9	25	138
May	2,081	56	52	187
June	2,028	54	53	282
July	4,508	15	130	370
August	6,930	24	272	403
September	2,410	29	135	219
October	981	24	22	53
TOTAL	**20,245**	**211**	**689**	**1,652**

Year 2017

Month	car	bus	Mini van	motorbike
April	479	2	16	17
May	2,091	42	67	172
June	2,159	48	79	247
July	5,147	27	201	398
August	7,468	29	310	422

(*Continued*)

Table CS5.1 (Cont.)

September	1,292	36	41	88
October	1,455	23	26	73
TOTAL	**20,091**	**207**	**740**	**1,417**

Source: Municipality of Solčava, 2020, available at: https://www.solcava.si/objava/261005
Note: 2020 data only covers the months from May–September inclusive

Table CS5.2 Analysis of visitors by transport mode in Logarska Valley (2019)

Mode of transport	Percentage of vehicles entering	Average annualised change 2017–2019	Entrance fee per person
Car	80.8%	1%	EUR2.34
Bus	13.6%	4%	EUR2.00
Motor-bike	4.7%	9%	EUR2.50
Mini-van/MPV	0.9%	16%	EUR1.60

Source: Authors analysis taken from Municipality of Solčava, 2020, available at: https://www.solcava.si/objava/261005

Note: no charge made for walkers or cyclists

Table CS5.3 Updated analysis for Logarska Valley (May–September 2020)

Mode of transport	Percentage of vehicles entering	Year-on-year change 2020/2019*
Car	90.4%	5.2%
Bus	0.1%	-82.7%
Motor-bike	6.4%	-10.6%
Mini-van/MPV	3.1%	-24.9%

Source: Authors analysis taken from Municipality of Solčava, 2020, available at: https://www.solcava.si/objava/261005

Note: no charge made for walkers or cyclists; * comparison of 2020 with 2019 based on May–September data only

EUR7.00 for a car, EUR8.00 for a minivan, EUR25.00 for a bus and EUR5.00 for a motor-bike. All funds collected are used for improvement to infrastructure and tourism standards in the valley, as well as matching funds when tendering for national and EU projects.

However, the COVID-19 pandemic had an impact on the Solčavsko and Logarska Valley areas. Up to 2019, on a peak day around 1,200 vehicles were registered entering the valley – based on average vehicle occupancies this suggest 3,600 visitors on a weekend peak day in August. By 2020 these figures were 50% down year-on-year implying 600 vehicles on an August peak day and thus 1,800 daily visitors. It was estimated in September 2020 that the annual flow of visitors for 2020 would be around 70–80% of previous years. However, it should be recognised that this represents a good result compared with the effect of COVID-19 on other tourist destinations.

Logarska Valley is viewed as offering a good agro-tourism product, primarily on a B&B basis. A good example of a popular destination is the Lenar Tourist Farm, which since 2010 has attracted foreign guests from Germany, Austria, UK, Israel and USA, as well as the Czech Republic, Poland and other countries in that region. In previous years during the June to August peak period, it was almost impossible to book a room, as guests were making repeat bookings prior to departure. However by May 2020 almost all foreign bookings were cancelled creating serious concerns on the part of owners. Fortunately, the introduction of the Slovenian government voucher programme triggered a strong level of bookings from the domestic market and as a result all capacities were sold out until mid-September 2020.

It appears that due to government intervention in stimulating the domestic market that the season has been saved with a resultant increase of income for 2020 at the beginning of the COVID-19 lockdown in March 2020. A further example is provided by the mountain lodge farm Roban on Robanov kot in the neighbouring Valley. This has been a very popular location for trekkers to stop and enjoy home-made traditional gastronomy specialities and drinks. Here, the owner reported, that despite fewer visitors in numbers, they saw good sales of their gastronomy products by domestic visitors who were familiar with these traditional food products, as opposed to foreign guests who would mostly take drinks only. This resulted in an approximate 25% of increase in profit, which is very good for the farm economy.

Similar experiences were noticed in other businesses around the valley proving that both local destination management in Solčava as well as the individual efforts of farm-tourism owners in Logarska Valley in promoting their business offer, and supported by state intervention in terms of the voucher programme, created a very positive and satisfactory outcome of what appeared likely to be a disastrous tourism season.

References

Anko, B., Anteric, M., Clarke, R., Koščak, M., Lenar, A., Mitchell, I., & Slapnik, M. (2007). *Študije o Solčavskem 1932–2007*. Poročilo o skupnem terenskem delu Univerze v Ljubljani in Univerze v Londonu. Solčava: Občina Solčava in Logarska dolina d.o.o., 118 p.

Lenar, A. (2013). Solčavsko, primer upravljanja lokalne skupnosti s krajinskim parkom. In: Bogataj, N. (ed.), *Znamenja trajnosti*. Ljubljana: ACS, 123–127.

Lenarčič, M., & Lenar, A. (2000). *Trajnostni razvoj v krajinskem parku Logarska dolina*. Nazarje: Argos.

Slapnik, M., & Bogataj, N. (2016). Nature- and Heritage-Based Tourism – Solčavsko as a Case of Resource Management by the Local Community. In: *Integral Green Slovenia, Towards a Social Knowledge and Value Based Society and Economy at the Heart of Europe*. Abingdon and New York: Routledge.

Other sources

Solčavsko. www.solcavsko.info (Accessed: 29 August 2020 and 8 September 2020)

Oral source: Lenar, A. (2020)

Case Study 6

Tourism voucher programmes – a post COVID economic dynamic?

Introduction

Since the efforts to release countries from lockdown in June 2020, resulting from the COVID-19 pandemic, much effort has been taken to seek to restore tourism and hospitality sectors which have been seriously affected by the crisis. Given that the release from lockdown has been affected by further spikes, there has been a general tendency in advanced economies to find means to stimulate domestic demand with a focus on tourism and hospitality.

This Case Study therefore seeks to examine:

1   Examples of methodologies utilised
2   The primary purpose of voucher schemes in seeking to boost demand for the tourist economy
3   Differences between programmes to incentivise the domestic stay-cation market and those to attract non-domestic tourists
4   The extent to which these schemes are strategically productive or simply short-term quick fix solutions

Please note that due to the varying levels of data available as at 1 November 2020, the statistical information applied by the authors is based on verbal data sources from interviews or extrapolations from public data.

Examples of methodologies utilised

The following are examples of systems established following the closing of global borders and the introduction of travel restrictions from April 2020.

AUSTRALIA

There are two examples of voucher schemes for domestic tourists in Australia intended to support tourism whilst there were a number of lockdowns on inter-state travel. Tasmania launched a "Make yourself at home" programme at the end of August 2020 to support Tasmanian businesses and the tourism sector. Under the AUD7.5m scheme families of five would be able to claim up to AUD550 in travel, accommodation and tourism experience costs from September 2020. The system allowed each adult to claim an AUD100 accommodation voucher and an AUD50 experience voucher, doubled for a couple and up to AUD550 for families with three children. However, on a first-come-first served system, once vouchers had been redeemed, citizens could apply for further vouchers as the scheme was planned to operate from early September through to 1 December. The Tasmanian Prime Minister suggested that the AUD7.5m input would have a AUD22.5m money multiplier effect. This is a cooperative effort between Tasmanian businesses and consumers, the Tasmanian Chamber of Commerce and Industry, the Tasmanian Government, Brand Tasmania and Tourism Tasmania. An accommodation voucher applied to establishments offering short-term stays on Sundays, Mondays, Tuesdays, Wednesdays or Thursdays during September, October or November 2020. Eligible property styles include apartments, backpacker and hostels, bed and breakfasts, cabins or caravans in holiday parks, cottages, farm stays, holiday houses, motels, hotels, resorts, retreats and lodges. Experience vouchers could also be used for tourism attractions, experiences and tours offered any day of the week.

In Australia's Northern Territory, also with sealed inter-state borders, the territory's government launched a Territory Tourism voucher for adults resident in the territory who would receive up to AUD200 if they contributed AUD200 of their own money. The whole programme would make up to 81,000 vouchers available over three rounds – Round 1, launched on 1 July 2020 was fully allocated; Round 2 was launched on 1 November 2020 and Round 3 will be launched on 1 February 2021. The total cost to the territorial government would be up to AUD15.2m. The vouchers can be used for bookable tourist products, and bookings can be made with listed tourism enterprises within 30 days of downloading for travel within four months of the launch of each of the voucher rounds.

STADT WIEN (VIENNA, AUSTRIA)

A EUR 50 voucher was given to residents of Stadt Wien (Vienna) to spend in local cafes and restaurants under the slogan "*Wie für Wiens wirte*". The gastro vouchers for city inhabitants were of a value up to EUR 50 and distributed to every household in the administrative area of Vienna. This provided an EUR 40m boost to help and support Viennese gastronomy after the lockdown and could be redeemed for food and non-alcoholic beverages at all participating restaurants and cafes.

SICILIA (SICLY, ITALY)

Given the dramatic drop in both visitors from abroad as well as domestic, Sicily's autonomous regional government offered to subsidise holidays for all visitors (domestic, European and international) by using an EUR 75m package of regional government funds. It is estimated that Sicily has lost around EUR 1bn in tourism revenue since the March 2020 shut-down (authors' estimates) so the package offered was a strong incentive to dynamise all tourism activity. The package included one night of a three night trip for free; in addition vouchers were provided for cultural and heritage activities – e.g. museums and archaeological sites. The discounts also applied to "non-hotel accommodation facilities" e.g. agro-tourism (farm stays), travel agencies, tour operators and registered tour guides. There have been suggestions from the regional government that the scheme may be extended as far as December 2021.

MALTA

All Maltese residents received EUR 100 vouchers to spend in bars, hotels and restaurants. The vouchers were issued in EUR 20 denominations – 4 vouchers could be used in hotels and restaurants and one voucher in bars. The cost to the Maltese government finances was estimated at EUR 900m.

PORTUGAL

On 25 April, as Portugal faced a massive downturn in tourism demand with 94% of hotels closed and 85% of workers on furlough schemes, the national tourism authority announced a delayed voucher plan for domestic, European and foreign tourists. The concept of the scheme was to encourage tourists whose vacations in Portugal had been cancelled due to flight and holiday package cancellations to apply for a voucher which would enable them to reschedule trips until the end of 2021.

The scheme applied to bookings through travel agencies or at accredited holiday accommodation (e.g. hotels or Airbnb properties) which had initially been scheduled to take place between 13 March and 30 September 2020. The vouchers were valid until 31 December 2021. Tourism sources reported that Portugal was acting as an absolute pioneer in the European context, whose priority was to safeguard consumer rights and the interests of economic operators, according to the principle of don't cancel, postpone.

SLOVENIA

In response to the Coronavirus pandemic the Slovenian government issued "Turistični boni" which are vouchers to every person who permanently resided in Slovenia on 13 March 2020. The vouchers are valid for qualifying accommodation with or without breakfast in Slovenia. Vouchers cannot be used to pay for other supplementary costs such as tourist tax or cleaning fees. Adults (over 18 in 2020) received a voucher for EUR200; those under 18 in 2020 received a voucher for EUR50. Vouchers are transferable between family members on a vertical level, i.e. between grandparents to grandchildren (but not at a horizontal level between brothers or sisters), are exempt from taxation and may be used at one time or over a number of visits. The vouchers were valid from 19 June 2020–31 December 2020 (with the possibility of being extended into 2021 if the situation in autumn 2020 proved the need for such measures).

Current statistics in August 2020 displayed that the Slovenian tourism season was rescued by the domestic market which was almost twice more in volume than in 2019. On the other hand, foreign guests only represented 30% compared to the numbers in previous season of 2019. The reasons for this are explicitly in regard to the introduction of the vouchers. The most visited types of accommodation were hotels on the coast (70% of visits were domestic market guests, almost all using vouchers), then apartments, camp sites all over Slovenia, agro-tourism with the most visited agro-tourism achieving around EUR40,000 from guests with vouchers as well as other visitors. The most negative tourism results so far were recorded in Ljubljana hotels – many of which were closed, whilst those which were open only recorded 14% of visits recorded in 2019. Ljubljana had been heavily visited by foreign guests in previous years, but apart from the effects of the pandemic there was a further reason for the reduction of foreign guests, namely the collapse of the Slovenian airline Adria Airways and as a result of COVID-19 the reduction of flights by other airliners into Ljubljana airport. It is interesting that for self-catering properties booked through the "Think Slovenia" website multi-week booking discounts have been in place offering a range of discounts for 1–2 week stays (5%) up to 25% on 6 week stays.

UK

Through August 2020, the UK government offered a scheme by which participating restaurants were able to offer meals at 50% of the actual costs subject to a maximum discount of GBP10. The programme was only applicable on Mondays to Wednesdays and the participating restaurants were able to submit claims for the cost of the discount to the UK tax agency. Initial data, as at 30 September 2020 (authors' analysis of UK provisional statistical material) indicates that 84,00 catering outlets (including hotels) participated in the programme and resulted in 64 million discounted meals. This resulted in 15% higher food sales than in August 2019 and it is estimated that footfall rose by 100% over Mondays, Tuesdays and Wednesdays in 2019. However, anecdotal comments indicated that Thursday became the new Monday, as large numbers of consumers fell away over the days the deal was not available. There is some evidence that catering outlets then continued, at their own cost, to continue offering discounts at a lower rate in order to maintain customer footfall. The UK-wide initiative did not discriminate towards the nationality of the consumer – as this covered domestic inhabitants as well as tourists from outside the UK.

The primary purpose of voucher schemes

BUSINESS SUPPORT

Voucher schemes are primarily a form of inducement to attract tourists by offering discounts through the application of accommodation vouchers or providing free access to important visit sites of a cultural or historical significance. It is clear that during the total lockdown periods in a number of countries, individuals were unable to travel and thus cultural and historical venues suffered severely from the lack of patronage. As lockdowns were eased, these venues were able to return to a form of structured access (e.g. pre-booking to limit numbers, social distancing at the venues, strict hygiene controls in the venue). To promote access levels to move swiftly towards the pre-COVID levels, many organisations have begun to offer free or discounted access pricing.

Similarly, accommodation facilities have also sought to offer discounted facilities on top of any national voucher schemes. It is very clear that renting houses, apartments or other self-contained accommodation (tents, glamping) has a huge hygiene advantage in that family or social bubbles are kept distinctly apart from other groups and from the potential for infection. Mass tourism, as we have seen from the August/September flows to Mediterranean resorts, has tended to bring increased levels of COVID-19 infection into the resorts, which were then augmented by the less than safe behaviour of under-29-year-old groups. This was then imported back into home destinations creating a number of COVID-19 spikes which began to be observed through August and September 2020.

Furthermore there has been strong domestic pressures in individual countries, regions and cities to encourage "staycationing". This suggests that individual families or family groups will select vacation experiences in their own countries or regions rather than taking the risk of travelling abroad and facing potential quarantine or infection risks through flights or other forms of mass travel.

DOMESTIC STAYCATION V NON-DOMESTIC TOURISTS

In many countries there has been a serious loss of international tourism arrivals, and this has resulted in vouchers schemes to attract domestic tourists. The Slovenian programme is an excellent example, as an estimated 50% of the Slovenian population have previously tended to take their summer vacations on the Croatian coast and islands. The variable situation resulting from COVID-19 infection in Croatia (border open – border closed) has therefore – with the voucher stimulus detailed above – resulted in larger numbers of Slovenes staycationing. Domestic voucher schemes or hospitality promotions such as the UK's "Eat Out – Eat Well" (as above) have also been helpful in boosting domestic demand and consumption.

At the same time, there are examples of programmes (such as in Sicily) to attract foreign visitors by offering discounts on accommodation and tourism visits. Unfortunately seeking to attract foreign tourists is highly dependent on the ability of those tourists to return from that destination without entering quarantine in their home country. Domestic tourism initiatives do not have this problem, unless there are areas of COVID shutdown, and therefore avoid the complexities of passing through airports with enhanced COVID testing.

Conclusion

Incentives provided by governments or other public authorities will tend to have a high profile in offering support to the tourism and hospitality sectors in terms of helping them cope with the loss of income sustained as a result of the COVID-19 pandemic.

To a certain extent they will be helpful in filling the revenue gap, especially where this relates to foreign tourism inflows. Equally, those projects aimed at domestic tourism will have the effect of boosting demand and supporting consumption of tourism and hospitality within the domestic market. A fundamental issue is that in general they are quick-fix programmes which generate good will towards state institutions but in the medium to longer term will not have any strategic consequences or effect.

Once the voucher scheme has expired, citizens may feel comfortable about the accommodation they have acquired and the meals they may have eaten, but will they be enticed to return to domestic vacations as a result of these incentives? There is little doubt that mass tourism vacation offers, currently suspended, will continue to attract tourists – it was evident in many Northern European markets that the opening of Spain and Portugal brought a massive surge of tourists to those destinations even though in a sense that surge may have contributed to the sudden spike of COVID-19 cases in the Iberian peninsular and resulted in the quarantining of Greece, Portugal and Spain from their key markets through July and August 2020. If you offer something for free or at a reduced price, will it really build loyalty or is it simply a short-term event to be taken advantage of?

Past events show us how fickle the mass tourism market may be; low-cost airlines offering flights at below EUR50 have tended to be advantageous for a relatively short period, but then the real cost hits the recipient airports and regions which host the flights. The process of low-cost flying to cater for short-term demands is also one of the issues which environmentalists have attacked – e.g. as in the Swedish inspired "*Flygskam*" (flight-shame) concept.

8   Analysis and conclusions

Introduction

At the time of writing, is not immediately apparent how long the pause will last for. After the first lockdown, it was optimistically believed that a subdued development of tourism globally would begin from the summer of 2020. This much heralded "new normal" failed to appear when from September/October 2020 not only did the developed economies (i.e. the members of OECD) begin to suffer rapidly expanding second waves, but the first wave continued to expand in a number of less developed economies for which tourism was a highly significant earner.

However, the key issue is undoubtedly that we cannot go back to where we were before COVID-19 materialised on the global tourism stage in early February 2020. It is clear that such types of pandemic are spread swiftly and efficiently by mass tourism transits – whether on cruise ships, inter-continental flights or high volume beach tourism. Prior to the pandemic, we had raised ethical and responsible tourism as an antidote to the environmental ravages of mass tourism on cultures, communities and physical environments. Importantly that must include the intangible as well as the tangible: "Policy tools to protect cultural heritage ... have traditionally focused on safeguarding physical assets. While this approach has often yielded the successful protection of historic sites, architecture and artefacts, it overlooks the intangible contributions" (Turner, 2020, p. 375). The policy tools appeared to be those of persuasion of the environmental and human benefits, backed by the ever nearing threat of climate change. The weapon that halted mass tourism and that has appeared to create a new reality in how we may replace it, has been the COVID-19 pandemic.

The strategic concerns

Trends

In pragmatic terms we cannot immediately persuade the millions who engage in mass tourism activities (e.g. high volume beach–sand–sea vacations or cruise ships) to immediately desist from those vacation opportunities based on the environmental damage that such vacations cause. It is unclear at present, for example, how influenced cruise ship passengers have been by the instances of passengers locked in cabins in February–March 2020 whilst the virus expanded at an exponential rate amongst them. In the past, the presence of Norovirus on cruise ships had not been a dampener for the huge growth in cruise tourism, but it was significantly less deadly. Mass tourism is relatively cheap for consumers and particularly profitable for operators, despite the fact that the mass tourism market may be easily influenced by negative trends (e.g. financial crises or terrorist acts against tourists). Mass tourism is to a certain extent an outcome of economic behaviour and consumer persuasion.

It has not been easy therefore to persuade consumers to adapt to a form of tourism which provides benefits to local populations that will:

- Expand sustainable, long-term employment opportunities
- Enhance work skills
- Preserve the natural, physical, cultural and historic heritage

Nonetheless, in the immediate aftermath of the first COVID-19 wave we observed tourism shifting towards domestic vacations and furthermore towards vacations in socially distanced accommodation (apartments, cottages, camping, glamping etc.). With barriers on travel and quarantine restrictions remaining in a constantly shifting number of countries, there was a move away from:

- Short-distance flights (e.g. up to four hours) due to the lack of social distancing
- Large accommodation facilities (e.g. large hotel or resort complexes and cruise ships) where there may be a lack of confidence in hygiene and medical safety
- Large transit locations (e.g. airports)

And there was a move towards vacations in sustainable destinations with a "slow tourism" effect and where there was a higher level of contact with local hosts and communities. The task ahead is to persuade consumers that such a type of travel is not only ethical and responsible, but may also be beneficial to a wider number of participants.

The timelines

With recovery paused – or in some cases moved even further forward into the future – the ability to reflect and then reset seems more distant. Indeed, coping with the short-term now appears likely to continue into 2023. Clearly some economies may potentially recover faster, i.e. in late 2021, but others will continue to be vulnerable and deeply traumatised well into 2023. Even in the advanced economies, which have capacity for mass vaccination programmes, full implementation of vaccination and an acceptable level of suppression of the COVID-19 virus is unlikely to take place until the autumn of 2021 and thus will affect the 2021–2022 winter tourism season. This would suggest that policies and plans for boosting ethical and responsible tourism capacity are required to commence from 2022–2025 three years of reflection and adaptation to a new and sustainable tourism future. Thereafter we may see a further nine year period for complete adaptation of the new norms, from 2025–2035.

Importantly it also requires moving away from the current tourism mindset-model which is to face any adverse event by flipping the consumer offer regardless of the effect on the host community. In 2015–2016, tourists to North African and Turkish resorts faced sporadic instances of terrorist attacks; airlines and tour operators simply switched capacity to European resorts with similar climatic conditions such as Greece, Cyprus, Spain and Malta. Local populations in Tunisia, Egypt and Turkey suddenly faced empty hotels, no jobs and no alternative capacity to general more sustainable forms of tourism.

Understanding a medium to longer-term strategic view

Despite the fact that the majority of tourism enterprises globally are in the form of SMEs (and in many cases, micro and small enterprises), the tourism industry as a whole is dominated by large companies. Whilst some are publicly quoted and relatively open, others are owned by hedge funds or venture capital organisations which are less transparent and less socially responsible. Problematically they are less oriented towards longer-term strategic thinking in a similar way to the large financial institutions to which they are connected. Thus the strategic concepts of these organisations is dictated by short-term capital flows and pricing concepts. In other words they operate on a very similar model to the failed banking institutions of the financial crisis of 2007–2008 and the subsequent economic depression of 2009–2015. The conditions which brought about the collapse of Thomas Cook in 2019 are not dissimilar to those which accompanied the collapse of a number of leading banks in 2008 – those who suffered most were initially customers as well as businesses reliant on these failed firms. Consumers were generally covered by state-backed guarantees covering holidays or bank deposits, but ultimately tax-payers were forced to carry the ultimate bill of those guarantees.

Ethical and responsible tourism, particularly at a local level does not carry such an intensive capital-risk-pricing package. We may see that local hosts and businesses are closer to their consumer and are less engaged by the variations in global markets. They are also more flexible in being able to adjust and redefine their offer, without dismissing employees and closing activities. Unfortunately they often lack strategic thinking and that is undoubtedly the element that must be maximised for them in order to expand and take advantage of a growing public appetite for sustainable tourism. It is evident that tourism with an ethical and responsible basis has already been able to engage the interests of three important groups (Koščak & O'Rourke, 2020):

- The "silver tourism market" – those above 60 years who are retired/semi-retired, remain physically active and have the financial assets to engage in independent organised vacations in places with significant cultural and heritage content
- The "family market" – families with children who are dissatisfied with mass tourism offers and seek a more authentic family experience which will gravitate towards active and adventure tourism
- The "back-packer market" – those who are between 20 and 30 years of age, seeking authentic local experiences, demonstrate a strong degree of flexibility in the travel plans and wish to engage with local communities

In the short to medium-term, over the next five years, all tourism activity has to undergo an adjustment to the new reality. Mass and mid-scale tourism will still be with us, but will be increasingly constrained by growing agendas by governments and societies for sustainable and environmentally supportable approaches. Ethical and responsible tourism will have the ability, if properly directed, to fill gaps in the market which will increasingly appear. Over the medium to longer term (i.e. 2023–2035), we must hope to see a greater role for ethical and responsible tourism on a more sustainable planet.

Tourism as an ecosystem

Overview

For many European and world regions and cities, tourism is a key contributor to the economic and social fabric. More than that, it provides much needed jobs and income, often concentrated in regions with no alternative sources of employment and involving low-skilled workers. However, this ecosystem has been hit hard with the United Nations World Tourism Organization (UNWTO, 2020) estimating a decline of international tourism of 65% for 2020 and of 51% for 2021 (both compared to 2019). This would project global revenue losses of USD 1.1 tn in 2020 alone.

Further, within the 27 countries of the European Union, tourism is the fourth largest EU revenue-earning category, which brings spill-over benefits to the EU27 economy as a whole. In effect EUR 1 of value added produced by tourism generates an additional EUR0.56 of value added in indirect effects on other industries.

Tourism is a complex ecosystem of many players: off-line and on-line information and services providers (tourist offices, digital platforms, travel technology providers), travel agents and tour operators, accommodation suppliers, destination managing organisations, visitor attractions and passenger transport activities. Tourism and transport are also built on major industries (e.g. construction, aircraft manufacture, shipbuilding). Large multinational corporations operate alongside small companies of which 80% are in the SME sector; tourism is present in many different types of areas and regions – urban centres, islands, coastal, rural and remote areas. Tourism is the backbone of the economy for many regions and destinations. Regions across Europe differ in terms of their reliance on tourism activities with strong impacts on islands, coastal and peripheral regions, due to their dependence on tourism activities or reliance on international air travel. Rural areas, with limited connectivity and dependence on tourism, have also felt a strong impact.

The impact of the crisis

The pandemic has placed the global and European tourism ecosystem under unprecedented pressure. As a result of travel and other restrictions, tourism reached a gradual halt during the first quarter of 2020 in the EU and globally. The European Commission (EU Commission, 2020) estimated the loss of economic growth in the five largest

European tourism economies at between 4.7% and 14.6%. According to the OECD (2020b) tourism is an economic sector which has been tremendously damaged by the coronavirus pandemic with an uncertain outlook for recovery. OECD estimated in October 2020 that international tourism in the world's most advanced economies will have negative growth in the area of 80% in 2020.

Tourism enterprises are facing an acute liquidity crisis. According to industry estimates, revenue losses at European level have reached 85% for hotels and restaurants, 85% for tour operators and travel agencies, 85% for long-distance rail and 90% for cruises and airlines. EU travel and tourism industry reports a reduction of bookings in the range of 60% to 90%, compared to the corresponding periods in previous years. The crisis has hit SMEs the hardest: lacking liquidity and facing uncertainty, they struggle to stay afloat, access funding, and maintain their employees and talent.

Without urgent action and emergency funding to bridge the period until limited tourism flows are likely to resume in spring 2021, many companies may become illiquid by the early months of 2021.

Jobs are also under threat. Tourism heavily relies on seasonal and temporary workers (23%), many of them young (37% of tourism workers are under 35), women (59%), and from other countries (15% EU or non-EU). These jobs are often concentrated in regions with no alternative sources of employment and involve low-skilled workers, with tourism accounting for 10% to 50% of total employment in many of them, including island, remote and outermost regions. Tourism also comprises an important section of social economy actors, contributing to social inclusion. Without urgent action to support employment, the crisis could lead to a loss around six million jobs in Europe and have a negative impact on the livelihood of many more people across different Member States, often among the most economically vulnerable (EU Commission, 2020).

The damage to tourism based economies

There is a general consensus that domestic tourism restarted from June/July 2020 and has had a critical role in supporting travel, tourism and hospitality across Europe, although this has been further affected by the growing series of lockdown and "circuit-breakers" which have occurred through the autumn of 2020. Indeed it is not unreasonable to assume that a return to growth will only take place when vaccination programmes have been fully implemented in the leading global economies in the autumn of 2021.

There is little doubt that the majority of governments in the advanced economies have generally worked to support tourism and the tourism hospitality and travel sector in general. But there remains a need to:

- Restore confidence in travel modes
- Support domestic tourism
- Seek to engage with alternatives to mass tourism which is extremely sensitive to global and economic changes
- Engage with the wider travel & tourism sector to see methods to maintain the industry in the short-term pending a return to moderate capacity in the medium term

OECD data indicated that prior to February 2020, tourism contributed 4.4% of GDP, 6.9% of employment, and 21.5% of service exports in the advanced economies which are members of the OECD. This average figure fails to recognise the critical role of tourism as a component of GDP in France (7.4%), Greece (6.8%), Iceland (8.6%), Mexico (8.7%), Portugal (8.0%) and Spain (11.8%). In these economies, where tourism underwent a second COVID wave shock, the economic effect has been substantially more significant.

Looking to the future

The short-term in Europe

Domestic and intra-EU tourism will prevail in the short-term. Some 267 million Europeans (62% of the population) make at least one private leisure trip per year and 78% of Europeans spend their holidays in their home country or another EU country (Eurostat, 2019).

Used creatively, once lockdown measures are lifted, the crisis offers an opportunity for Europeans to enjoy the rich diversity of culture and nature in their own or other EU countries and discover new experiences all year around.

Many European regions and cities rely heavily on cultural tourism. Cultural tourism, representing 40% of tourism in Europe, particularly has suffered as most cultural activities, like fairs and festivals, were cancelled and institutions such as museums closed (92%). Technology has helped to reinvent cultural tourism during this pandemic by opening new opportunities for creative expression and by expanding audiences. Furthermore, it may be estimated that over half of European tourism capacity is located in coastal areas and almost one third of tourism nights are spent at beach resorts. Whilst high-capacity non-sustainable coastal tourism may operate on an all-year round basis, smaller rural/regional resorts, which are significantly more sustainable and ethically responsible, will be required in future to explore off-season capacities, e.g. niche tourism modes as well as small scale meetings, conferences and events. New opportunities are arising to discover hidden or forgotten natural and cultural gems closer to home and to taste locally produced products.

In several Member States where patronage voucher schemes have been set up, customers have shown enthusiasm for supporting their favourite hotels or restaurants. Such schemes could extend to other tourism related businesses such as culture and entertainment. This could be showcased on an IT portal which would link up suppliers with all initiatives and platforms offering such schemes. It would help customers to find all initiatives and offers throughout the EU. In addition and in cooperation with many Member States and their regional and local tourism destinations, the EU Commission could call for pledges to launch patronage voucher systems from local tourism organisations, but also from market players who are active in the tourism sector, such as small and large online platforms (through which many tourism businesses connect with their clients), credit card companies and payment system providers.

Implications in advanced economies

Data from OECD (2020a) implies that:

1   Sustainability will potentially take a greater significance as a result of the collapse of mass tourism as well as a heightened understanding by tourism consumers of climate change and the adverse impacts of tourism
2   This would suggest that local destinations which have strong environmental and sustainable content will drive tourism recovery and will be based on tourism which has a lower environmental impact
3   We would also suggest that domestic tourism will gain a greater share, as tourists will in general be more reluctant to risk the health and hygiene threats of long-distance travel. For example, travel by over-night sleeping car trains which not only provide a discrete personal environment but also are mask-free, may grow in volume over air travel.
4   At the same time, we should keep in mind that due to the collapse of mass tourism, a large number of skilled tourism employees will

be released onto the labour market. Many of them will be unable to find employment in locally focused tourism activities and may therefore shift to other economic activities. The new reality may be that the jobs these individuals once had will no longer exist

5   We also must recognise the fact that the global investment sector continues to support large-scale tourism activity due to its immense equity and debt exposure. At the same time, global investment has shown scant regard to engagement with small and micro-scale tourism activity, despite the actual inherent risk balance being consistently lower.

6   This will grow the need for locally based investment and funding, which in essence will release a currently unseen capacity at a local and regional level. But this will require more flexible and SMART financing technologies, including blending loan/equity/mezzanine financing as well as crowd funding and the use of small-scale micro-equity markets

New horizons – towards a sustainable future

Beyond the immediate steps to bring relief, we should look ahead to the future of tourism and transport in the EU and look at how to make it more resilient and sustainable, learning from the crisis and anticipating new trends and consumer patterns related to it. Our shared ambition should be to maintain Europe as the world's leading tourist destination in terms of value, quality, sustainability and innovation. This vision should guide the use of financial resources and investment at European, national, regional and local levels.

At the core of this new ambition is sustainability and responsibility. A joint aim should be to enable affordable and more sustainable transport and improved connectivity, boost smart management of tourism flows based on sound measurement and tools, diversify the tourism offer and extend off-season opportunities, develop sustainability skills for tourism professionals, and valorise the variety of landscapes and the cultural diversity across Europe – while protecting and restoring Europe's land and marine natural capital, in line with the strategic approach for a sustainable blue and green economy. This should include the promotion of sustainable tourism accommodations and other innovative schemes such as EU Ecolabel and the Eco-Management and Audit Scheme (EMAS). This ambition towards sustainable tourism should guide investment decisions at EU level, but it can only work in combination with a strong commitment at regional and local level (EU Commission, 2020).

Equally, tourism can gain from the digital transition, providing new ways of managing travel and tourists flows, opportunities, and more choice, as well as more efficient use of scarce resources. The use of big data analysis can create and share accurate tourist profile segments and help understand traveller trends and needs. It can enable tourism to respond immediately to the changing customer demand and provide predictive modelling analysis. Finally, the application of blockchain technology would allow tourism operators to have all the available information about safe transactions.

Digital tools can also be confidence-building measures to reassure people that travel and tourism can be safe. This means investment in digital skills, including cybersecurity and fostering digital innovation, and connecting tourism businesses and actors with existing data spaces at local and regional level (for instance through the ongoing work on the European data space on mobility). This is particularly important in rural, remote areas and outermost regions where tourism is fragmented and highly dependent on information, transport, and travel accessibility.

Within this digital transition, micro and small enterprises (MSEs) will need particular attention in supporting such tourism MSEs become more resilient and competitive. This requires building cross-sectoral linkages, interdisciplinary knowledge flow, stronger connections and capacity to ensure accelerated uptake of sustainable digitalised products, services and process innovations. Such networks should connect tourism with other industries to accelerate uptake of new solutions whilst fostering cross-sectoral investments in the tourism ecosystem with ICT, renewable energies, health and life sciences, agri-food, marine, the media, cultural and creative sectors.

New sustainable trends?

The sobriety in 2020 for the tourism sector was probably strong and painful enough, unfortunately, the potential for further shocks to occur until mass-vaccination occurs is inevitable. However, the experience of 2020 shows that the crisis has been successfully survived by those destinations that provide individual safety and distance, quality natural environment, remoteness from the masses, etc. Before 2020, the key words for the development of sustainable tourism (Koščak & O'Rourke, 2020) could be seen as related to:

- Responsible management
- Seeking added value for local products and services
- Safety and social responsibility

- Accessibility for all
- Assessment of the carrying capacity of the environment
- Managing excessive tourism
- Participatory planning and partnership approach

Certainly this remains important for the future development of sustainable tourism. However, especially in view of the events in 2020 which have not yet seen their end, the following new trends and implementation of measures that will ensure the following in tourist destinations should also be viewed as important in our new reality:

- Experiential approach in product design and new experiences to travellers
- Security
- General and personal hygiene
- Social distancing
- The environmentally sustainable approach towards a preserved environment
- Conditions for the well-being of guests
- Addressing nearer markets

We believe that many local destinations will be able to provide all the above elements of the new reality. Without doubt those who are successful in that process will have to face the task of transforming their development strategies and searching for new development paradigms which intensively incorporate the principles of sustainable and responsible tourism. Diluting of those principles will result in sliding back towards the complacency which marked the tourism sector prior to 2020. For local tourism to enter a fully sustainable and responsible environment and to develop and flourish within such an environment, requires a fundamental commitment to enduring change over the medium to longer term.

References

EU Commission (2020) Communication from the Commission to the European Parliament, the Council, the European Economic and Social Committee and the Committee of the Regions, Tourism and transport in 2020 and beyond. Retrieved 13 November 2020 from https://ec.europa.eu/info/live-work-travel-eu/health/coronavirus-response/jobs-and-economy-during-coronavirus-pandemic_en#supporting-recovery-of-eu-tourism.

Eurostat (2018) Characteristics of employment in tourism, EU-28, 2017. https://ec. europa.eu/eurostat/statistics-explained/index.php?title=Tourism_industries_-_ employment#Characteristics_of_jobs_in_tourism_industries.

Eurostat (2019) 'People on the move. Statistics on mobility in Europe',

Koščak, M., & O'Rourke, T. (eds) (2020) *Ethical and Responsible Tourism: Managing Sustainability in Local Tourism Destinations.* Abingdon, Oxon, New York: Routledge.

OECD (2020a) Making the green recovery work for jobs, income and growth. OECD Paris, 6 October.

OECD (2020b) Rebuilding tourism for the future: COVID-19 policy response and recovery. OECD, Paris, 22 October.

Turner, B. (2020) Promoting the "Legacy Businesses" of San Francisco, Chapter 27, 375–383 in Koščak, M., & O'Rourke, T. (eds): *Ethical and Responsible Tourism: Managing Sustainability in Local Tourism Destinations.* Abingdon, Oxon, New York: Routledge.

UNWTO (May 2020) International tourist numbers could fall 60–80% in 2020, UNWTO reports. Retrieved May 15, 2020 from https://www.unwto.org/news/covid-19-international-tourist-numbers-could-fall-60-80-in-2020.

Index

active tourism 43 *see also* slow tourism
adventure tourism 10, 136

bottom-up approach 4, 10, 50
business adaptation 6, 52

carrying capacity 62, 70, 83–84, 118, 143; and carrying capacity study and processes: 4, 12, 50; footfall 10
Clemente, A.A. xi
climate change 6, 19, 20, 53, 56, 133, 140
community-led development 4, 50, 52; and community transport 72
COVID-19 pandemic 132–135; and economic effects: 27–31; financial effects 41–44; local tourism: 69, 98, 110,124–126; major economies 31–35; mass tourism 63–65; pandemic development 1, 11; tourism damage **14, 17,** 48–51; transformation of tourism 20, 25
crisis and disaster 14, 66
cross-sectoral 142
cultural tourism 43, 139

destination manangement 56–57, 70–71, 94, 125; and concepts: 3–4; impacts 49–51; opportunities 21–22
Dubrovnik, case study 82–90

eco-tourism 72, 74
environmental protection 92
European Tourism Council 76, 78

ethical tourism 56, 108, 109 *see also* responsible tourism
ethnography 24
European Union 31, 113, 137
experience economy 22, 71
experiential travel 4

financing ethical & responsible tourism 6, 7, 8, 34, 52–53
fragile environments 6, 10, 55, 80–81

gastronomy 49, 93, 98, 125, 128
globalisation 30, 47, 64
governance 68, 76, 121

health, safety and security 57, 58, 60, 66, 67 *see also* COVID-19 pandemic
heritage tourism 4, 5, 8, 9, 10–12, 117 and intangible heritage: 10, 133; local heritage structures: 120; regulation 61; strategy and trends 133–134, 137; sustainability 72–74
Host communities 56

International Monetary Fund 28–29, **32, 33,** 36

landscapes 4, 50, 92, 96, 141
local level 4, 6, 10, 24–25, 41, 56, 117 and communities: 51–52, 82, 136; economics 43–44; development 72–73; local perspectives 68–78; local structures 8, 9, 140 *see also* community-led development

Logarska Dolina, case study 109–117

Malta, case study 109–117
managing sustainability 7, 11, 48, 54, 98, 144
market changes 7, 53
mass tourism 55–56, 63–65, 71, 81, 96, 139–140 and economic recovery: 44–45, 78; environmental impact108–109, 116, 122; health and safety 21, 23, 24 *see also* overtourism
micro and small enterprises 43, 45, 135, 142 *see also* financing ethical and responsible tourism

natural environment 5, 68, 81, 119, 129, 139, 142
negative impacts 5, 61, 82
networking 6, 52
new reality 2, 11, 37, 80–81 and elements of new reality: 36, 39–40, 136, 141, 143

Organisation for Economic Cooperation & Development 34–35, 40–41, 138–140
overtourism 9, 25, 51, 112 *see also* mass tourism

participatory planning 4, 50, 120, 143
pause-reflect-reset 81
peripheral regions 7, 8, 137
public-private-social partnerships 4, 7, 12 50, 91, 143

renewable energy 6
resilience 43–44, 67, 70, 77–78

responsible tourism 6, 7, 48, 50, 56–58, 136 and consumptive behaviour 33, 41, 81, 92, 132; critical factors 8–12; responsible consumption 27, 31, 39, 73, 75 *see also* ethical tourism
regeneration 8, 81

slow tourism 71, 134 *see also* active tourism
SMART technology 7, 53
social distancing 35–36, 38, 39, 89, 114, 131, 134, 143 *see also* health, safety and security
social responsibility 5, 51, 142
sustainability 11, 48–49, 50, 70, 140 and sustainable tourism: 19, 24, 39–40, 62, 68–69

tourism development 11, 49–53, 70 and tourism planning: 50, 61; tourism recovery from COVID-19 36, 37, 38, 39
tourism voucher programmes, case study 126–132
travel restrictions 13, 16, 49, 76–77, 127

UNESCO 5–6, 87, 99
UNWTO 15–18, 35, 82, 99, 137

Venice, case study 99–109

waste management 73
wine tourism 81, 92–93 and wine tourism accommodation 90–91, 94, 97
wine tourism, case study 90–98

9780367716318